T0187064

Eduqas
Biology
for A Level
Yr1 & AS

Revision Workbook

Neil Roberts

Published in 2020 by Illuminate Publishing Limited, an imprint of Hodder Education, an Hachette UK Company, Carmelite House, 50 Victoria Embankment, London EC4Y 0DZ

Orders: please contact Hachette UK Distribution, Hely Hutchinson Centre, Milton Road, Didcot, Oxfordshire, OX11 7HH. Telephone: +44 (0)1235 827827. Email: education@hachette.co.uk. Lines are open from 9 a.m. to 5 p.m., Monday to Friday. You can also order through our website: www.hoddereducation.co.uk

British Library Cataloguing in Publication Data

A catalogue record for this book is available from the British Library

ISBN 978-1-912820-39-9

Printed by Ashford Colour Press, UK

Impression 4
Year 2024

Hachette UK's policy is to use papers that are natural, renewable and recyclable products and made from wood grown in well-managed forests and other controlled sources. The logging and manufacturing processes are expected to conform to the environment regulations of the country of origin.

Publisher: Adrian Moss

Editor: Geoff Tuttle

Design and layout: Nigel Harriss

Cover image: iStock.com/vallesiles

Acknowledgements

For Isla and Lucie.
The author would also like to thank the editorial team at Illuminate Publishing for their support and guidance.

About the author

Dr Neil Roberts PhD FRSB, is a former head of Biology and has over 20 years' teaching experience in universities, schools and colleges in England and Wales, and was an experienced principal examiner for a major awarding body. He is a fellow of the Royal Society of Biology.

Picture Credits

p32, p34 ©Dr Neil Roberts
p114, Shutterstock.com/Nelson Bastidas
p116, Dr Keith Wheeler/Science Photo Library
p132, Biophoto Associcates/Science Photo Library
p137, Shutterstock.com/BioMedical

other illustrations ©Illuminate Publishing

Contents

How to use this book

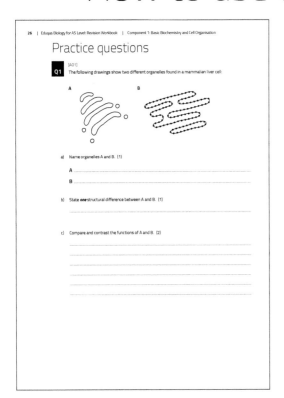

Each topic section of the guide begins with around four to eight examination-style practice questions focussed on each topic. The assessment objective being tested is identified for you. The real exam questions will frequently draw from several topics, so it is important, once you feel confident with the practice questions, you move on to complete past papers (one is provided for each section in this book) and many are available from the exam board website.

This is then followed by some examples of actual student answers to questions. In each case there are two answers given; one from a student (Isla) who achieved a high grade and one from a student who achieved a lower grade (Ceri). We suggest that you compare the answers of the two candidates carefully; make sure you understand why one answer is better than the other. In this way you will improve your approach to answering questions.

Examination scripts are graded on the performance of the candidate across the whole paper and not on individual questions; examiners see many examples of good answers in otherwise low-scoring scripts. The moral of this is that good examination technique can boost the grades of candidates at all levels.

This exam preparation guide has been designed to work alongside the Study and Revision Guide, also available by the same author and published by Illuminate Publishing.

Assessment objectives

Examination questions are written to reflect the assessment objectives (AOs) as laid out in the specification. The three main skills that you must develop are:

AO1: Demonstrate knowledge and understanding of scientific ideas, processes, techniques and procedures.

AO2: Apply knowledge and understanding of scientific ideas, processes, techniques and procedures.

AO3: Analyse, interpret and evaluate scientific information, ideas and evidence, including in relation to issues.

In both written examinations you will also be assessed on your:
● Mathematical skills (minimum of 10%)
● Practical skills (minimum of 15%)
● Ability to select, organise and communicate information and ideas coherently using appropriate scientific conventions and vocabulary.

These are indicated against each topic question together with mathematical skills (M) and practical skills (P) where they occur.

In any one question, you are likely to be assessed on all skills to some degree. It is important to remember that only about a third of the marks are for direct recall of facts. You will need to apply your knowledge, too. If this is something you find hard, practise as many past paper questions as you can. Many examples come up in slightly different forms from one year to another.

Your practical skills will be developed during class-time sessions and will be assessed in the examination papers. This could include:
● Plotting graphs
● Identifying controlled variables and suggesting appropriate control experiments
● Analysing data and drawing conclusions
● Evaluating methods and procedures and suggesting improvements.

Understanding AO1: Demonstrate knowledge and understanding

You will need to demonstrate knowledge and understanding of scientific ideas, processes, techniques and procedures; 35% of the questions set on the exam paper are for recall of knowledge and understanding.

Common command words used here are: state, name, describe, explain. This involves recall of ideas, processes, techniques and procedures detailed in the specification. This is content you should know.

A good answer is one that uses detailed biological terminology accurately and has both clarity and coherence. If you were asked to describe the structure of starch and explain how its structure makes it a good molecule for storage of glucose, you might write:

'Starch is made from alpha glucose molecules joined together. Because it is insoluble it is ideal for storage.'

This is a basic answer.

A good answer needs to be both accurate and detailed. For example:

'Starch is a polymer of alpha glucose molecules joined by condensation reactions. It consists of two molecules: amylose, which is a straight chain formed from 1-4 glycosidic bonds; and amylopectin, which is highly branched, formed from alpha glucose joined by 1-4 glycosidic bonds with 1-6 branches every 20 or so glucose molecules. Starch is very compact, and because amylose and amylopectin are both insoluble it does not affect the water potential of the cell, and so makes an ideal storage molecule in plant cells. Amylopectin releases glucose very quickly as there are many ends on which the enzyme amylase can act.'

Understanding AO2: Applying knowledge and understanding

You will need to apply knowledge and understanding of scientific ideas, processes, techniques and procedures:

- In a theoretical context
- In a practical context
- When handling qualitative data (this is data with no numerical value, e.g. a colour change)
- When handling quantitative data (this is data with a numerical value, e.g. mass/g).

Approximately 45% of the questions set on the exam paper are for application of knowledge and understanding. Common command words used here are: describe (if its unfamiliar data or diagrams), explain and suggest. This involves applying ideas, processes, techniques and procedures detailed in the specification to unfamiliar situations.

Describing data

It is important to describe accurately what you see, and to quote data in your answer.

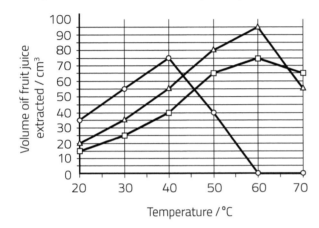

- ⊸ Free enzymes
- ⊶ Enzyme bound to gel membrane surface
- ⊟ Enzyme immobilised inside beads

If you were asked to compare the volume of juice produced when using enzymes bound to the gel membrane surface compared to the enzyme immobilised inside the beads, you might write:

'The volume of juice extracted increases with temperature up to the optimum temperature of 60 °C in both enzymes. Above this, the volume of juice decreases.'

This is a basic answer.

A good answer needs to be both accurate and detailed. For example:

'Increasing temperature causes the volume of fruit juice extracted to increase up to 60 °C. The volume of juice collected is higher up to 60 °C with the enzyme bound to the gel membrane, peaking at 95 cm³ compared to 75 cm³ for the enzyme immobilised inside the beads. Above 60 °C the volume of fruit juice extracted decreases, but this is more noticeable for the enzymes bound to the gel membrane surface which decrease by 40 cm³ compared to just 10 cm³ for the enzyme immobilised inside the beads.'

If you were also asked to explain the results, a basic answer would include reference to *'increased kinetic energy up to 60 °C, and denaturing enzymes above 60 °C'*. A good answer is one that uses detailed biological terminology accurately and has clarity and coherence. A good answer would also include reference to *'increased enzyme–substrate complexes forming up to 60 °C'* and would include that *'above 60 °C, hydrogen bonds break, resulting in the active site changing shape so fewer enzyme–substrate complexes could form'*.

Mathematical requirements

A minimum of 10% of marks across the whole qualification will involve mathematical content. Some of the mathematical content requires the use of a calculator, which is allowed in the exam. In the specification, it states that calculations of the mean, median, mode and range may be required, as well as percentages, fractions and ratios. Some additional requirements are included at A level, which are shown in a bold italic font. You will be required to process and analyse data using appropriate mathematical skills. This could involve considering margins of error, accuracy and precision of data.

Concepts	Tick here when you are confident you understand this concept
Arithmetic and numerical computation	
Convert between units, e.g. mm^3 to cm^3	
Use an appropriate number of decimal places in calculations, e.g. for a mean	
Use ratios, fractions and percentages, e.g. calculate percentage yields, surface area to volume ratio	
Estimate results	
Use calculators to find and use power, exponential and logarithmic functions, e.g. estimate the number of bacteria grown over a certain length of time	
Handling data	
Use an appropriate number of significant figures	
Find arithmetic means	
Construct and interpret frequency tables and diagrams, bar charts and histograms	
Understand the principles of sampling as applied to scientific data, e.g. use Simpson's Diversity Index to calculate the biodiversity of a habitat	
Understand the terms mean, median and mode, e.g. calculate or compare the mean, median and mode of a set of data, e.g. height/mass/size of a group of organisms	
Use a scatter diagram to identify a correlation between two variables, e.g. the effect of lifestyle factors on health	
Make order of magnitude calculations, e.g. use and manipulate the magnification formula: magnification = size of image / size of real object	
Understand measures of dispersion, including standard deviation and range	
Identify uncertainties in measurements and use simple techniques to determine uncertainty when data are combined, e.g. calculate percentage error where there are uncertainties in measurement	
Algebra	
Understand and use the symbols: =, <, <<, >>, >, ∝,~.	
Rearrange an equation	
Substitute numerical values into algebraic equations	
Solve algebraic equations, e.g. solve equations in a biological context, e.g. cardiac output = stroke volume × heart rate	
Use a logarithmic scale in the context of microbiology, e.g. growth rate of a microorganism such as yeast	
Graphs	
Plot two variables from experimental or other data, e.g. select an appropriate format for presenting data	
Understand that $y = mx + c$ represents a linear relationship	
Determine the intercept of a graph, e.g. read off an intercept point from a graph, e.g. compensation point in plants	
Calculate rate of change from a graph showing a linear relationship, e.g. calculate a rate from a graph, e.g. rate of transpiration	
Draw and use the slope of a tangent to a curve as a measure of rate of change	
Geometry and trigonometry	
Calculate the circumferences, surface areas and volumes of regular shapes, e.g. calculate the surface area or volume of a cell	

Understanding AO3:
Analysing, interpreting and evaluating scientific information

This is the last and most difficult skill. You will need to analyse, interpret and evaluate scientific information, ideas and evidence, to:
- Make judgements and reach conclusions
- Develop and refine practical design and procedures.

Approximately 20% of the questions set on the EDUQAS exam paper are for analysing, interpreting and evaluating scientific information. Common command words used here are: evaluate, suggest, justify and analyse. This could involve:
- Commenting on experimental design and evaluating scientific methods
- Evaluating results and drawing conclusions with reference to measurement, uncertainties and errors.

What is accuracy?

Accuracy relates to the apparatus used: How precise is it? What is the percentage error? For example, a 5ml measuring cylinder is accurate to ±0.1ml so measuring 5ml could yield 4.9–5.1ml. Measuring the same volume in a 25ml measuring cylinder which is accurate to ±1ml would yield 4–6ml.

Calculating % error

It's a simple equation: accuracy/starting amount ×100. For example, in the 25ml measuring cylinder the accuracy is ±1ml so the error is 1/25 ×100 = 4%, whereas in the 5ml cylinder the accuracy is ±0.1ml so the error is 0.1/5 ×100 = 2%. Therefore, for measuring 5ml it is better to use the smaller cylinder as the % error is lower.

What is reliability?

Reliability relates to your repeats. In other words, if you repeat the experiment three times and the values obtained are very similar, then it indicates that your individual readings are reliable. You can increase reliability by ensuring that all variables that could influence the experiment are controlled, and that the method is consistent.

Describing improvements

If you were asked to describe what improvements could be made to the reliability of the results obtained from an experiment extracting apple juice, you would need to look closely at the method and apparatus used.

Q: Pectin is a structural polysaccharide found in the cell walls of plant cells and in the middle lamella between cells, where it helps to bind cells together. Pectinases are enzymes that are routinely used in industry to increase the volume and clarity of fruit juice extracted from apples. The enzyme is immobilised onto the surface of a gel membrane, which is then placed inside a column. Apple pulp is added at the top, and juice is collected at the bottom. The process is shown in the diagram. Describe what improvements could be made.

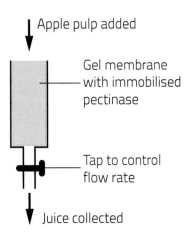

You might write:

'I would make sure that the same mass of apples is added, and that they were the same age.'

This is a basic answer.

A good answer needs to be both accurate and detailed. For example:

'I would make sure that the same mass of apples is added, for example 100g, and that they were the same age, e.g. 1 week old. I would also control the temperature at an optimum for the pectinases involved, e.g. 30°C.'

Look at the following example:

A student carried out an experiment to investigate the effect of temperature on cell membranes. Using a borer, equal sized pieces of beetroot were cut, washed, and blotted with a paper towel. Each piece was placed into a test tube containing 25 cm³ of 70% ethanol (an organic solvent) and incubated at 15°C. A red pigment called betacyanin found in the vacuoles of the beetroot cells began to leak out into the ethanol turning it red. The experiment was repeated at 30°C and 45°C and the time taken for the ethanol to turn red was recorded in the table below:

Temperature (°C)	Time taken for the ethanol to turn red (s)			
	Trial 1	Trial 2	Trial 3	Mean
15	450	427	466	447.7
30	322	299	367	329.3
45	170	99	215	161.3

Q: What conclusions could be drawn from this experiment regarding the effect of temperature on cell membranes?

You might write:

'Increasing temperature increases the amount of dye that leaks from the cells.'

A good answer needs to be both accurate and detailed. For example:

'Increasing temperature increases the kinetic energy of the membrane and dye molecules. The increased movement of membrane molecules increases the number of gaps in the membrane so more dye can escape from the cells.'

If asked to comment on the validity of your conclusion, you might write:

'It was difficult to determine when the solutions turned red, making it difficult to know when to stop timing the reactions.'

A good answer would be more detailed. For example:

'The results at 45°C are very variable and range from 99 to 215 seconds. It is difficult to reach a conclusion about the effect of temperature on cell membranes as only three temperatures were investigated. Another major difficulty would be in determining the end point of the reaction, as no standard red colour was used.'

Preparing for the examination

Types of exam question

There are two main types of questions in the exam.

1. Short-answer structured questions

The majority of questions fall into this category. These questions may require description, explanation, application, and/or evaluation, and are generally worth 6–10 marks. Application questions could require you to use your knowledge in an unfamiliar context or to explain experimental data. The questions are broken down into smaller parts, e.g. a), b), c), etc., which can include some 1-mark name or state questions. You could also be asked to complete a table, label or draw a diagram, plot a graph, or perform a mathematical calculation.

Some examples requiring name or state:
- Name the bond labelled X on the diagram. (1 mark)
- Name the type of cell division taking place. (1 mark)
- What is the name given to this type of diagram? (1 mark)

Some examples requiring mathematical calculation:
- The magnification of the image above is × 32 500. Calculate the actual width of the organelle in micrometres between points A and B. (2 marks)
- Using the formula and the table given below, calculate the Diversity index. (3 marks)

Some examples requiring description:
- Describe the results for the free enzymes at temperatures above 40°C. (2 marks)
- Describe one similarity and one difference between the structures of chitin and cellulose. (2 marks)
- Describe how a sweep net could be used to estimate the Diversity index of insects at the base of a hedge. (3 marks)

Some examples requiring explanation:
- Explain why there must be three bases in each codon to assemble the correct amino acid. (2 marks)
- Giving examples, explain the difference between homologous and analogous structures. (2 marks)
- Explain how the structures of cellulose and chitin are different from that of starch. (2 marks)

Some examples requiring application:
- Using your knowledge of the structure of cell membranes, explain why ethanol causes the red pigment betacyanin to leak out of the beetroot cells. (2 marks)
- Explain why the rate of water uptake in the plant increased as the wind speed increased. (3 marks)
- What conclusions could be drawn from this experiment regarding the effect of temperature on cell membranes? (3 marks)

Some examples requiring evaluation:
- Describe how you could improve your confidence in your conclusion. (2 marks)
- Comment on the validity of your conclusion. (2 marks)
- Evaluate the strength of their evidence and hence the validity of their conclusion. (4 marks)

2. Longer essay questions

As part of the exam, you will need to answer an extended response question worth 9 marks. The quality of your extended response (QER) will be assessed in this question. You will be awarded marks based upon a series of descriptors: to gain the top marks it is important to give a full and detailed account including a detailed explanation. You should use scientific terminology and vocabulary accurately, including accurate spelling and use of grammar and include only relevant information. It is a good idea to do a brief plan before you start to organise your thoughts: You should cross this out once you have finished. We will look at some examples later.

Command or action words

These tell you what you need to do. Examples include:

Analyse means to examine the structure of data, graphs or information. A good tip is to look for trends and patterns, and maximum and minimum values.

Calculate is to determine the amount of something mathematically. It is really important to show your working (if you don't get the correct answer you can still pick up marks for your working).

Choose is to select from a range of alternatives.

Compare involves you identifying similarities and differences between two things. It is important when detailing similarities and differences that you discuss both. A good idea is to make two statements, linked with the word 'whereas'.

Complete means to add the required information.

Consider is to review information and make a decision.

Describe means give an account of what something is like. If you have to describe the trend in some data or in a graph then give values.

Discuss involves presenting the key points.

Distinguish involves you identifying differences between two things.

Draw is to produce a diagram of something.

Estimate is to roughly calculate or judge the value of something.

Evaluate involves making a judgment from available data, conclusion or method, and proposing a balanced argument with evidence to support it.

Explain means give an account and use your biological knowledge to give reasons why.

Identify is to recognise something and be able to say what it is.

Justify is about you providing an argument in favour of something; for example, you could be asked if the data support a conclusion. You should then give reasons why the data support the conclusion given.

Label is to provide names or information on a table, diagram or graph.

Outline is to set out the main characteristics.

Name means identify using a recognised technical term. Often a one-word answer.

State means give a brief explanation.

Suggest involves you providing a sensible idea. It is not straight recall, but more about applying your knowledge.

General exam tips

Always read the question carefully: read the question twice! It is easy to provide the wrong answer if you don't give what the question is asking for. All the information provided in the question is there to help you to answer it. The wording has been discussed at length by examiners to ensure that it is as clear as possible.

Look at the number of marks available. A good rule is to make at least one different point for each mark available. So make five different points if you can for a four-mark question to be safe. Make sure that you keep checking that you are actually answering the question that has been asked – it is easy to drift off topic!

If a diagram helps, include it: but make sure it is fully annotated.

Timing

You have 90 minutes, so this gives you a sense of how much time you should spend on each exam question, and a good rule is about one mark per minute. Don't forget that this timing is not just about writing but you should spend time thinking, and for the extended answer some planning, too.

Common exam mistakes

1. **Misreading the question!**
   Sounds obvious I know but READ the question carefully – know your command words!

2. **Not including enough detail**
   You should budget on a minimum of one mark a point, so if the question is out of 5, make at least 5 points, making sure to include your biological knowledge.

3. **Using incorrect terminology**
   Reliability is NOT the same as accuracy!

4. **In maths answers ALWAYS show all your working**
   Credit is given if the examiner can see your steps even if you do end up with the wrong answer, so always show your working in full AND remember your units. If you don't you could lose a mark!

5. **Spelling errors**
   Spelling of key scientific words MUST be right; for example, examiners won't accept 'meitosis' for meiosis because it could be confused with mitosis! The general rule is examiners will allow phonetic spelling so long as it can't be confused for something else. However, for the extended question the quality of your extended response IS assessed.

6. **Describing data**
   Remember to quote data from graphs/tables in your answer WITH units.

7. **Running out of time**
   Exam papers are written to give you plenty of time. If you get stuck, move on BUT remember to come back to it later. Every year marks are lost because some questions are left blank!

8. **Drawing graphs**
   Full marks are rarely awarded for graphs. Common errors include:

   - Incorrect labels on axes
   - Missing units
   - Sloppy plotting of points
   - Failing to join plots accurately
   - Non-linear scales.

Have a go yourself – can you spot the mistakes?

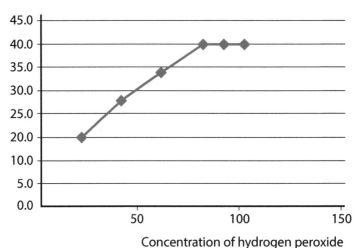

Mistakes are:

- No units on either axis.
- No value for origin on horizontal axis.
- Vertical axis is nonlinear, i.e. gaps are unequal.

Also make sure that you draw range bars and can explain their significance.

Component 1: Basic Biochemistry and Cell Organisation

Section 1: Chemical elements and biological compounds

Topic summary

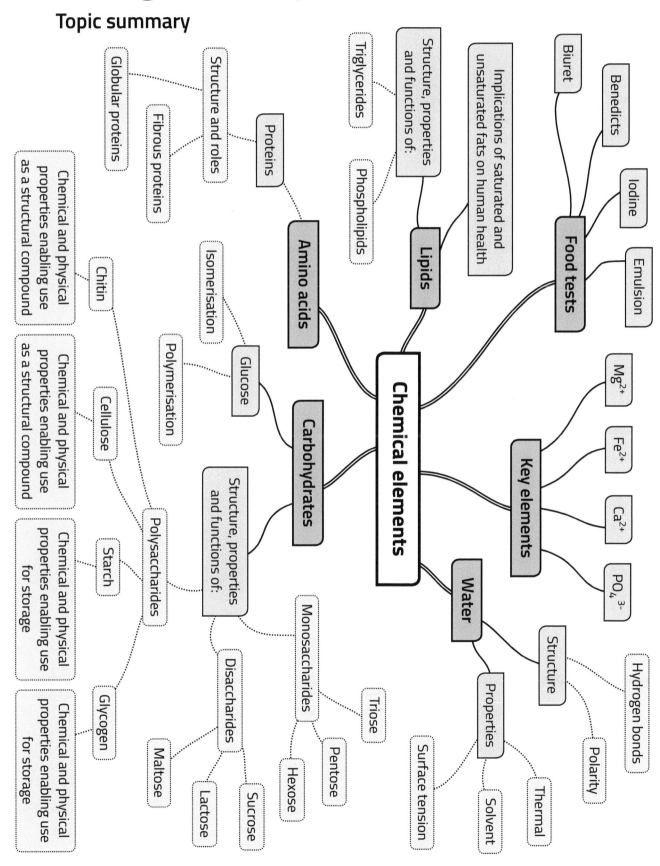

Practice questions

EXAM TIP

You need to know the structure of the elements as this affects how they join together and their properties.

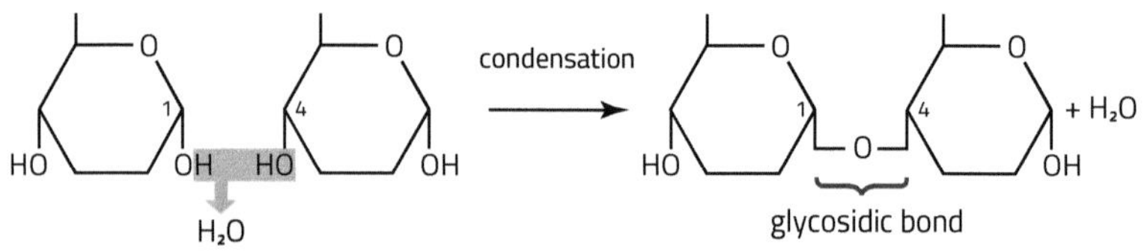

Formation of a glycosidic bond between two glucose molecules making maltose

EXAM TIP

You need to be able to draw diagrams to show how two units join together by a condensation reaction, and how they are split by hydrolysis.

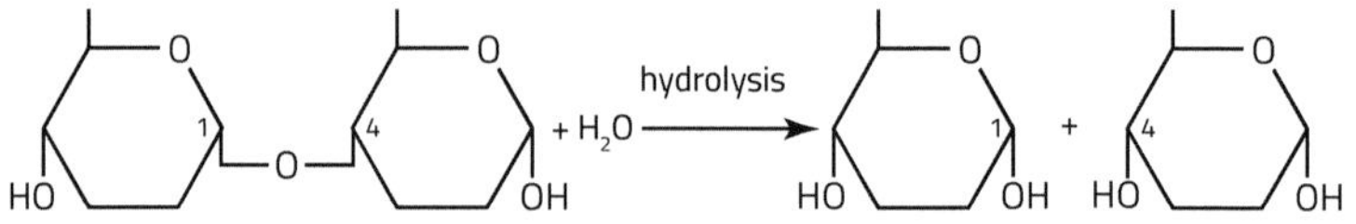

Hydrolysis of glycosidic bond in maltose

[AO1]

Q1 The following diagram shows the monosaccharides glucose and galactose.

a) Complete the diagram to show how the molecule is joined to form a disaccharide. (2)

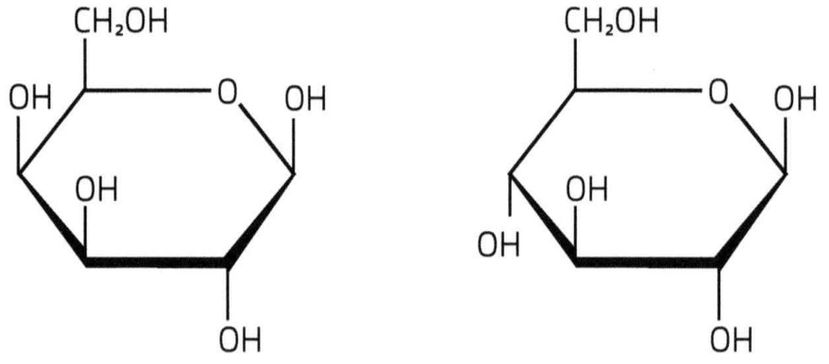

b) Name the reaction. (1)

....

c) Name the disaccharide produced and the bond formed. (2)

....

d) Glucose is a reducing sugar. Explain what is meant by a reducing sugar. (1)

....

Q2

[a,b = AO1] [c = AO2]

The diagram below shows two water molecules joined by bond A.

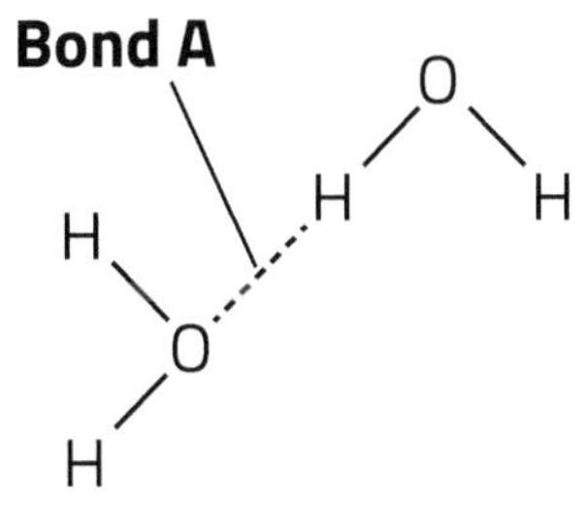

Bond A

a) Name bond A. (1)

....

b) Annotate the diagram to show the charges present on the oxygen and hydrogen atoms. (1)

c) Explain how the charges you have identified in part b) enable water to be drawn up the xylem vessels of trees. (3)

....

....

....

....

....

[AO1]

Q3 The following diagrams show sections of two polysaccharides:

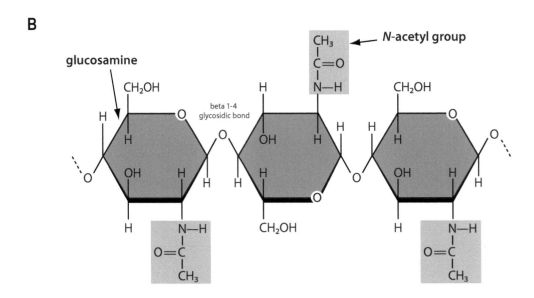

a) Name both polysaccharides shown. (1)

A ....

B ....

b) Compare the structure and function of both A and B. (5)

....

....

....

....

....

....

....

....

[AO1]

Q4 The diagram below shows two different molecules.

A

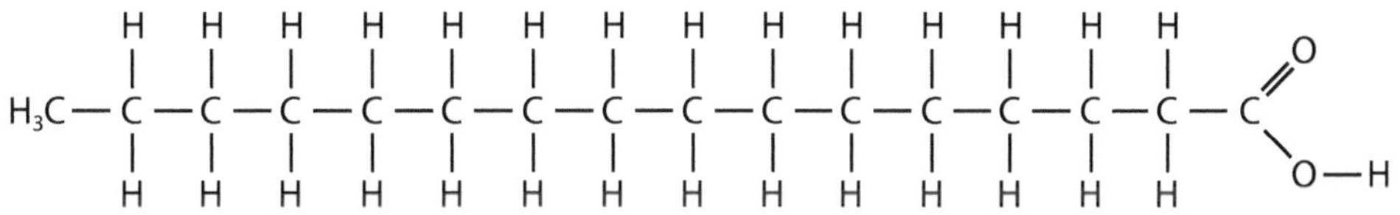

B

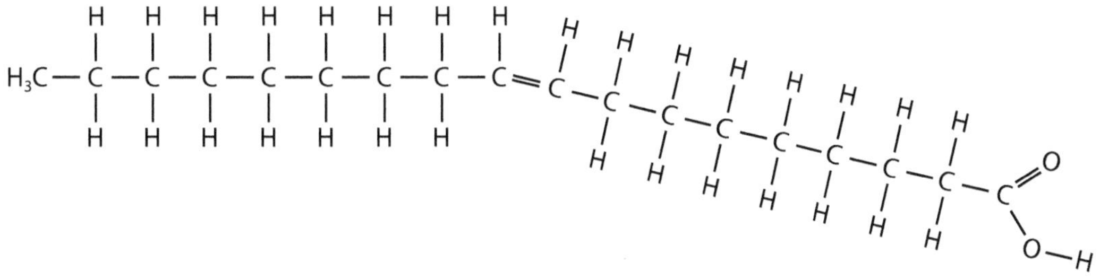

a) Which group of biological molecules do they belong to? (1)

....

b) Draw a circle on diagram A to indicate the molecules removed during a condensation reaction with a glycerol molecule. (1)

c) Explain how the structure of molecule A differs from molecule B and so influences its properties. (3)

....

....

....

....

....

[a = AO1, b = AO2]

Q5 The following diagram shows a dipeptide:

Bond A

a) Draw a diagram in the space below to show how the dipeptide is broken down into its constituent molecules by hydrolysis. Name bond A broken. (3)

b) There are about 20 different amino acids used to make up proteins, each differing in their composition of the R group. An experiment determined the percentage composition of different molecules, and the peak activity of three different enzymes. These are shown in the table below:

Enzyme	Peak enzyme activity /°C	Percentage composition / %				
		Carbon	Hydrogen	Oxygen	Nitrogen	Sulphur
A	37	15.0	34.2	25.2	15.6	0.0
B	45	18.6	36.4	24.8	18.2	2.0
C	55	17.0	28.2	22.8	13.4	18.6

Scientists concluded that enzyme C would be the most tolerant to high temperatures. Explain how the experimental evidence above supports this conclusion. (4)

EXAM TIP

You need to use data from the table AND your biological knowledge to explain the conclusion.

Q6 A student tested three different solutions, A, B and C, using Benedict's reagent, and biuret reagent. The colours obtained following each test are shown in the table below:

Solution	Colour result with	
	Benedict's reagent	Biuret reagent
A	Brick red	Blue
B	Blue	Purple
C	Green	Purple

a) What conclusions can be made about the contents of solutions A, B and C? (3)

Solution A ....

Solution B ....

Solution C ....

b) A fresh sample of solution B was boiled with dilute hydrochloric acid, neutralised and then heated with Benedict's reagent. The result was a brick-red colour.

Give an explanation for this result. (2)

....

....

....

c) Describe how you could use biuret reagent to determine whether solution B or C contained the highest concentration of protein. (4)

....

....

....

....

....

....

Component 1 Practice questions

[a = AO1] [b, c = AO3]

Q7 Mass spectrometry and size-exclusion chromatography were used to measure the length in nanometres (nm) of the primary and secondary structures of a large polypeptide at 20°C and at 60°C. The results are shown in the table below:

Structure of polypeptide	Length of polypeptide / nm	
	At 20°C	At 60°C
Primary	495	495
Secondary	45	73

a) Explain what is meant by the term secondary structure of a protein. (2)

....

....

....

....

b) State *two* conclusions that can be made from the results shown. (3)

....

....

....

....

....

c) Suggest reasons for your conclusions above. (3)

....

....

....

....

....

....

Q8

[AO1]

Collagen and haemoglobin are both proteins.

a) State what is meant by the primary structure of a protein. (1)

....

b) Name *one* type of bond other than peptide bonds that is present in both molecules. (1)

....

c) Complete the table to show *five* comparative *differences* between collagen and haemoglobin. The first one has been done for you. (3)

Collagen	Haemoglobin
Only made of polypeptide chains	Made of polypeptide chains and a prosthetic haem group
Fibrous protein	
Structural protein	
Consists of one type of polypeptide chain	

Question and mock answer analysis

[AO1]

Q&A 1 The following diagram shows part of a polysaccharide:

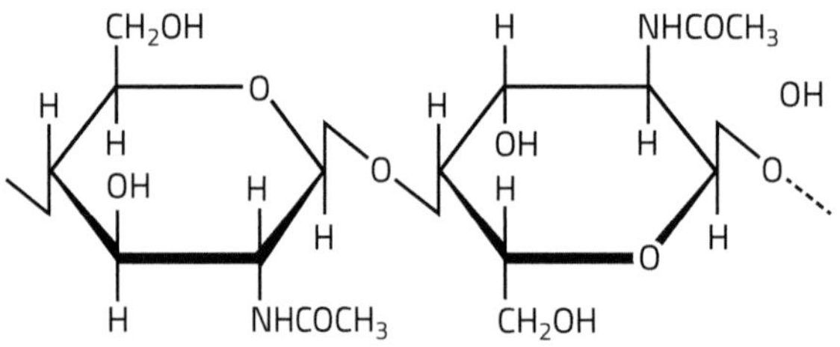

EXAM TIP

It is important that you read the question carefully and give as detailed a response as you can. Spelling is important, especially if you write a word that could be confused with something else, in this case glucose. When you are asked to give a difference, you must say something about both – in this case cellulose and chitin.

a) Name the monomer that makes up the polymer, and its form. (1)

b) Name the bond formed between the two hexose sugars. (1)

c) State one structural difference between this molecule and cellulose. (1)

d) Explain how the molecule shown gives strength to the exoskeleton. (2)

Isla's answers

a) β glucose. ✓

b) 1–4 glycosidic bond. ✓

c) some OH groups have been replaced with NHCOCH₃ ✓

d) hydrogen bonds form between the straight chains of β glucose molecules which then form microfibrils. ✓

MARKER NOTE

Isla's answer should have been expanded to to include that adjacent molecules are rotated by 180° and that hydrogen bonds form between parallel chains.

Isla achieves 4/5 marks

Ceri's answers

a) glucose ✗

MARKER NOTE
Ceri has not named the form, β.

b) 1–4 glucosidic bond ✗

MARKER NOTE
Ceri has misspelled glycosidic which could be confused with other terms.

c) there are NHCOCH₃ present ✗

MARKER NOTE
Ceri has identified the additional group but has not compared it with cellulose by saying that the OH groups have been replaced.

d) chains form microfibrils ✗

MARKER NOTE
Ceri did not explain how the chains form microfibrils.

Ceri achieves 0/5 marks

Q&A 2

[AO1]

Using examples, explain how the structure of carbohydrates and lipids enables them to perform their variety of functions in living organisms. (9 QER)

EXAM TIP

It is not one mark per point in the extended response, but more about how you answer the question. A plan is essential if you are to write coherently, and you must use scientific terminology correctly.

Isla's answer

Carbohydrates contain carbon, hydrogen and oxygen and are used in respiration, as storage molecules and provide structural support. Glucose is a six-carbon monosaccharide, which is the main source of energy for living organisms. It is easily hydrolysed during respiration producing ATP, ✓ and acts as a building block for more complex polysaccharides. Starch is used to store glucose in plants because unlike glucose it is insoluble and therefore osmotically inert. ✓ It is made from amylose and amylopectin.

> **MARKER NOTE**
> Isla describes the main functions of glucose well.

Amylose contains alpha glucose molecules joined via 1–4 glycosidic bonds into straight chains which twist to form helices. In amylopectin the alpha glucose molecules are more branched, due to 1–4 and 1–6 glycosidic bonds. This creates a structure that is highly compact and so is easily stored within plant cells ✓ but it is easily hydrolysed into glucose when needed for respiration.

> **MARKER NOTE**
> The structure of starch has been described well to explain its role as a storage molecule.

In animals, glucose is stored in the form of glycogen. Glycogen is another highly compact molecule that is osmotically inert, but it is made from a branched molecule containing alpha glucose joined by 1–6 glycosidic bonds, similar in structure to amylopectin. ✓ Plants and animals also use carbohydrates for structural support. Plant cell walls are strengthened by cellulose, which is made from chains of beta glucose molecules. Alternating glucose molecules rotate 180° and form straight chains. ✓ Hydrogen bonds then form between the long parallel chains forming microfibrils, which in turn are held together into fibres. ✓ The presence of fibres arranged at right angles to other fibres in the cell wall provides strength. ✓ In insects, chitin is formed in a similar way, except some OH groups are replaced by acetylamine groups. Their arrangement is similar, resulting in a strong and lightweight molecule found in the exoskeleton of insects. ✓

> **MARKER NOTE**
> The formation of microfibrils is clearly described and used to explain the role of chitin in structural support.

Lipids also contain carbon, hydrogen and oxygen, but the proportion of oxygen is less. They are non-polar molecules and hence insoluble in water, enabling their use as a waterproofing agent in the form of leaf waxes, and oils on birds' feathers. ✓ Their insolubility also makes lipids a good energy store in the form of oils in seeds, and saturated fatty acids in animals. ✓ Storing fats under the surface of the skin also helps prevents heat loss as fats are poor conductors of heat, ✓ and as a component of myelin, they surround neurones, providing electrical insulation. ✓ Due to the high numbers of hydrogen atoms, their hydrolysis releases more energy per gram than carbohydrates – in fact twice the energy, ✓ and also releases metabolic water, which is important in desert animals like the kangaroo rat where water is scarce. Some fats are stored around delicate organs such as the kidney, providing some protection against physical damage. Lipids are also key components in the plasma membrane of cells where they form phospholipids. ✓

> **MARKER NOTE**
> The properties of lipids have been used to explain their roles.

Carbohydrates and lipids are also needed to make other molecules important to plants and animals such as nucleotides with the addition of phosphate ✓ and chlorophyll with magnesium. ✓

EXAMINER COMMENTARY

Isla would receive full marks (9/9) for an answer that fully addresses the question using good examples to link structure to function of lipids and carbohydrates with no irrelevant inclusions or significant omissions. Isla uses scientific conventions and vocabulary appropriately and accurately.

Component 1 Practice questions

Ceri's answer

Carbohydrates are used in respiration, and as storage molecules and provide structural support. Starch is stored in plants because unlike glucose it is insoluble and compact and so is easily stored within plant cells. ✓

In animals, glucose is stored in the form of glycogen. Plant cell walls contain cellulose, which is made from beta glucose molecules which are made into fibres in the cell wall provides strength.

Lipids are insoluble in water, so make good energy in animals.

They have more energy than carbohydrates and also release water, which is important in deserts. Fats are stored around delicate organs like the kidney providing protection.

They also are found in the plasma membrane of cells where they form phospholipids.

MARKER NOTE

Ceri could have included that starch has no effect on osmosis in cells. Ceri would need to mention that starch is an energy/glucose store, and that the helical and branched structures of amylose and amylopectin make it more compact.

MARKER NOTE

It is not clear exactly what provides strength here. Ceri could have included detail on how beta glucose molecules are arranged into chains and microfibrils, and details on chitin.

MARKER NOTE

Whilst lipid insolubility is mentioned, Ceri could link it to providing waterproofing.

MARKER NOTE

Need to say twice as much energy.

MARKER NOTE

Ceri needs to clarify what protection is conferred, i.e. protection from physical damage.

EXAMINER COMMENTARY

Ceri would receive 1 out of 9 marks because whilst some relevant points are made, showing limited reasoning, the structure of the key elements enabling them to perform their function is not explained. Ceri has made limited use of scientific conventions and vocabulary.

Section 2: Cell structure and organisation

Topic summary

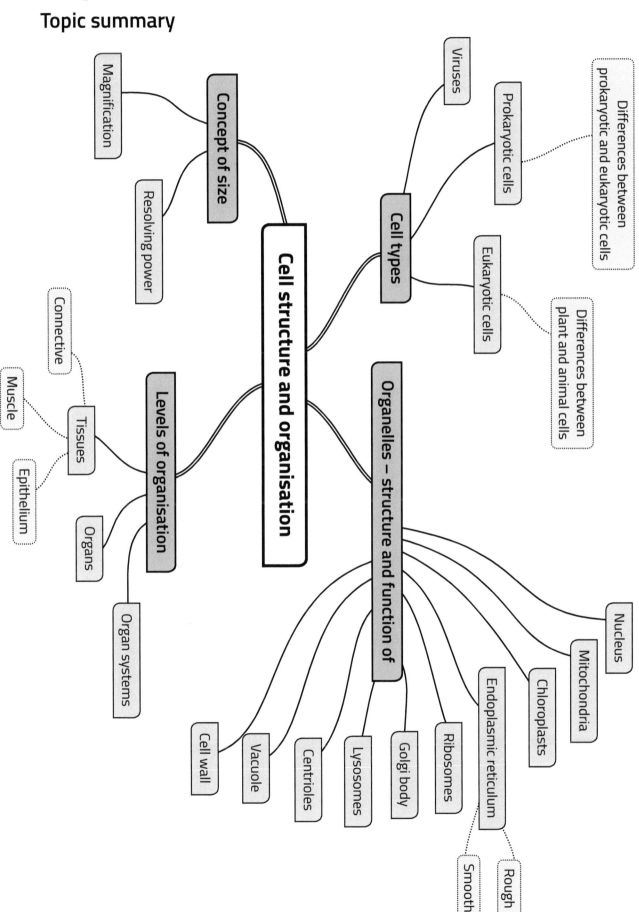

Practice questions

 Q1

[AO1]

The following drawings show two different organelles found in a mammalian liver cell:

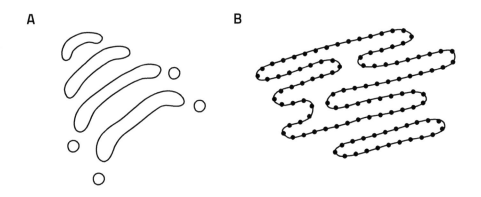

A   B

a) Name organelles A and B. (1)

A ....

B ....

b) State *one* structural difference between A and B. (1)

....

c) Compare and contrast the functions of A and B. (2)

....

....

....

....

....

....

[AO1] [d = AO2]

Q2 The following drawings were made of cells from two different epithelial tissues when seen under the light microscope:

A

B

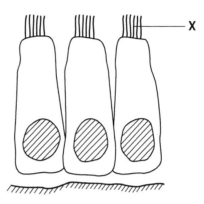

a) Name **two** features present in **both** tissues. (1)

....

b) Identify tissue B and suggest a place in the human body where it might be found. (2)

B tissue

....

B found

....

c) Name feature **X** and explain how its function adapts the tissue for where it is found. (2)

....

....

....

....

....

[AO1]

Q3 a) Complete the following table to compare typical animal and prokaryotic cells. (3).

Typical animal cell	Typical prokaryotic cell
No cell wall	
No chloroplasts	No chloroplasts
	70 S ribosomes
	Ribosomes loose in cytoplasm
Mitochondria	
Nucleolus	No nucleolus
	May contain plasmids

b) Draw a diagram below to show the basic structure of a ribosome found in prokaryotic cells, labelling the mRNA and tRNA attachment sites and subunits. (3)

Q4

[a = AO1] [b, c = AO3]

10g of liver tissue was taken and homogenised by grinding the sample in a blender with 50 cm³ of cold isotonic buffer. The extract was then centrifuged at 5000 rpm and the pellet collected. The extract was centrifuged again at increasing speeds until no further pellets were obtained. This process is known as cell fractionation.

The results are shown in the table below:

Centrifuge speed (rpm)	Organelles collected
5,000	Nuclei
7,000	Mitochondria
10,000	Lysosomes and vesicles
15,000	Ribosomes

a)   Explain why cold isotonic buffer was used. (2)

....

....

b)   What can you conclude about the size of the organelles? (2)

....

....

....

....

c)   Outline a simple experiment you could perform to confirm that mitochondria were collected. (2)

....

....

....

....

....

[AO1]

Q5 The following diagram shows a general prokaryotic cell:

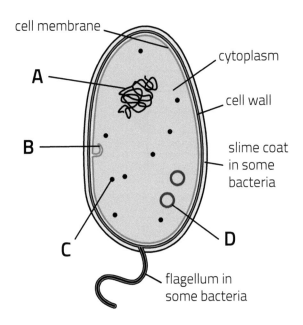

a) Complete the table below to identify the feature and its function. (4)

Feature	Name	Function
A		
B		
C		
D		

b) Outline how feature C would differ in an eukaryotic cell. (2)

....

....

....

....

[AO1]

Q6   The following diagram shows an organelle from a eukaryotic cell:

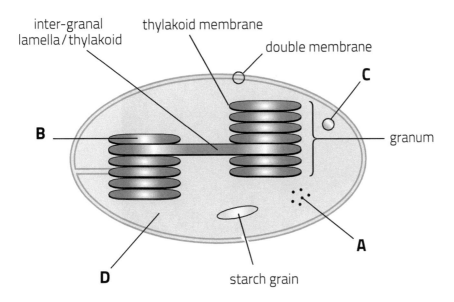

a)   Name the organelle shown. (1)

....

b)   Complete the table below to identify the feature and its function. (4)

Feature	Name	Function
A		
B		
C		
D		

c)   Describe how the organelle shown above differs from a mitochondrion. (2)

....

....

....

....

[P] [AO1]

Q7 The photograph below shows the head of a parasitic wasp *Chlorocytus* species, taken using a scanning electron microscope, which has a resolving power of about 10 nm.

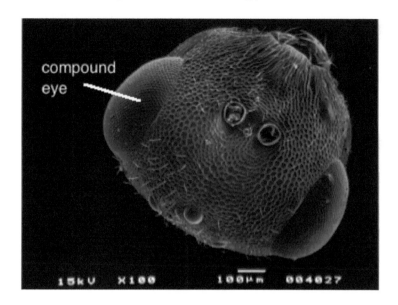

a) The photograph shows two compound eyes, which are regarded as organs. State what is meant by the term organ. (2)

....

....

....

....

b) What is the difference between resolving power and magnification? (2)

....

....

....

....

[AO1]

Q8   a)   In the space below draw a labelled diagram to show a Golgi body. (3)

b)   Describe how the structure and function of the Golgi body differ from rough endoplasmic reticulum. (4)

....

....

....

....

....

....

....

....

Question and mock answer analysis

[a = AO1, b = AO2, c = AO3]

Q&A 1 The following electron micrograph shows an organelle found in muscle tissue:

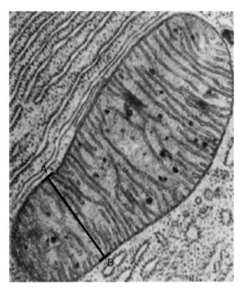

a) The nucleus has pores in the envelope that surrounds it, whereas the organelle shown does not. Name the organelle shown and describe one other difference between the membranes that surround the organelle shown and the nucleus. (2)

b) Estimate the surface area of the organelle shown using the formula surface area $= 2\pi r l + 2\pi r^2$, where l = length of the organelle is 9.8 µm, π = 3.14, diameter is 0.9 µm. Show your working. (3)

c) The surface area of a spherical organelle with the same volume is 25% less. Results from experiments have concluded that there are advantages to the cell of these organelles possessing a cylindrical shape. Evaluate this statement. (4)

Isla's answers

a) Mitochondrion. ✓
   Mitochondria have a folded inner membrane which the nucleus doesn't. ✓

b) Radius = 0.5 ✓

   $(2 \times 3.14 \times 0.45 \times 9.8) + (2 \times 3.14 \times 0.20)$ ✓

   55.39 + 1.26

   = 56.65 (µm²) ✓

> **MARKER NOTE**
> Isla needs to include the mechanism by which materials are exchanged, e.g. diffusion, and make reference to diffusion distance.

c) If a spherical organelle's surface area is 25% less than a cylindrical one for the same volume, less substance, e.g. oxygen, can be exchanged across the surface to meet the needs of the organelle. ✓ A cylindrical mitochondrion will have a smaller distance to the centre than a spherical one ✓ because its radius is smaller, and so it will take less time for oxygen to reach the innermost parts of the organelle and for carbon dioxide to be lost, which is an advantage to cells respiring. ✓

Isla achieves 8/9 marks

> **EXAM TIP**
> It is really important to link the structure to its function in these sorts of questions.

Component 1 Practice questions

Ceri's answers

a) Mitochondrion. ✓
   hey have a folded inner membrane.

MARKER NOTE
Ceri did not compare both organelles.

b) Radius = 0.5 ✓ = 56.64 (µm²)

MARKER NOTE
Ceri did not show any working, and incorrectly rounded the answer – ALWAYS show your working so credit can be given.

c) A spherical mitochondrion will have a larger diffusion distance to the centre than a cylindrical one ✓ which results in more oxygen being absorbed.

MARKER NOTE
Ceri failed to describe the method of absorption and needed to discuss the SA:Volume ratio which is lower in a spherical mitochondrion.

Ceri achieves 3/9 marks

[AO1]

Q&A 2 Cell walls are important features in many cells.

a) Describe how the basic composition of the cell wall differs in plants, fungi and prokaryotes. (3)

b) Explain how the structure of the cell wall enables it to function in plants. (6)

Isla's answers

a) In plants it is made from the polysaccharide cellulose, ✓ in fungi it is chitin ✓ and in prokaryotes it is peptidoglycan. ✓

b) One function is to provide strength to the cell wall and support ✓ in non-woody plants.

MARKER NOTE
Isla has not explained how the wall provides support, i.e. the wall resists the expansion of the cell contents and creates turgor.

The cell wall is also involved in transport ✓ as water and dissolved molecules and ions pass through gaps in the cellulose fibres via the apoplast pathway. ✓ Communication between cells ✓ is possible via pores in the cell wall allow strands of cytoplasm called plasmodesmata to pass. ✓ This also allows water to pass via the symplast pathway.

Isla achieves 8/9 marks

Ceri's answers

a) plants – cellulose ✓
   fungi – chitin ✓
   prokaryotes – peptidoglycan. ✓

b) The cell wall provides strength and support in non-woody plants. ✓

MARKER NOTE
Ceri has also failed to reference resisting the expansion of cell contents creating turgor. This is a common omission.

c) It also helps with transport as water can pass through gaps in the walls,

MARKER NOTE
Not gaps in the walls but gaps between fibres. Gaps would refer to plasmodesmata.

and with communication between cells ✓ because it has plasmodesmata which allows water to pass via the symplast pathway. ✓

MARKER NOTE
Ceri has stated three transport mechanisms but only explained how one of them works.

Ceri achieves 6/9 marks

Section 3: Cell membranes and transport

Topic summary

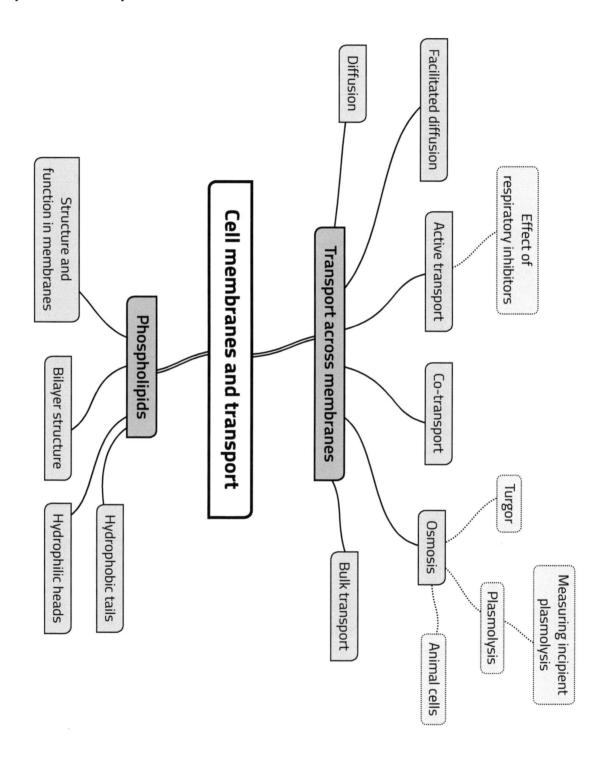

Practice questions

[AO1 / AO2]

Q1 The diagram below shows a section of the cell membrane:

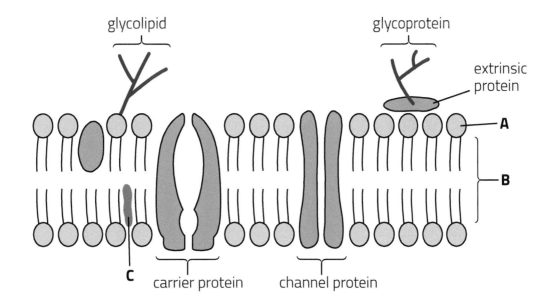

a) Name the following components: (2)

A ....

B ....

C ....

b) Explain how features shown in the diagram assist in the passive movement of polar molecules across the membrane. (3)

....

....

....

....

....

[P] [M] [AO3] [c = AO2]

Q2 An experiment was carried out to investigate the effect of oxygen concentration on the uptake of oxygen into cells at 20°C and 40°C. The results are shown below:

Concentration of oxygen (kPa)	Rate of uptake (a.u.) at 20°C	Rate of uptake (a.u.) at 40°C
0	0.0	0.0
3	2.5	3.9
6	5.5	7.4
9	8.9	9.9
12	12.1	13.9

a) Draw a line graph below to show the effect of increasing oxygen concentration on the rate of uptake. (5)

b)   Use your graph and data to draw conclusions. (5)

....

....

....

....

....

....

....

....

c)   Explain how you could modify the experiment to show that uptake was not by active transport. (2)

....

....

....

....

....

[M] [P] [AO2, d = AO3]

Q3 A student carried out an experiment to estimate the solute potential of onion cells by placing a piece of onion epidermis onto a glass slide and covering with 0.1 cm³ of sucrose solution ranging from 0.0M to 0.5M. After 30 minutes the slide was viewed under a light microscope and the number of cells showing plasmolysis recorded.

a) Complete the table below to show percentage plasmolysis. (2)

Molarity of sucrose solution (M)	Number of cells plasmolysed	Total number of cells observed	Percentage plasmolysis
0.0	0	70	
0.1	19	75	
0.2	28	65	
0.3	37	66	
0.4	49	72	
0.5	65	71	

b) Draw a graph to find incipient plasmolysis. (5)

Incipient plasmolysis = ... (1)

c) Using the conversion table below, estimate the solute potential of the onion epidermal cells. (1)

Conversion table

Molarity of sucrose solution (M)	Solute potential kPa
0.05	−130
0.10	−260
0.15	−410
0.20	−540
0.25	−680
0.30	−860
0.35	−970
0.4	−1120
0.45	−1280
0.50	−1450
0.55	−1620
0.60	−1800

Answer

....

d) What can be concluded about the water potential, solute potential and pressure potentials of the onion epidermal cells at incipient plasmolysis? (2)

....

....

....

....

....

Component 1 Practice questions

[a = AO2, b = AO3]

Q4 Scientists investigating the co-transport of glucose in the ileum measured the concentration of sodium ions and glucose in the lumen of the ileum, epithelial cells lining the ileum, and the blood. The results are shown below:

Molecule	Concentration (mmol / L^{-1}) in		
	the lumen of the ileum	epithelial cells lining the ileum	the blood
Sodium ions	83.9	4.9	130.0
Glucose	12.7	42.0	5.1

a) Using your knowledge of transport mechanisms and the data in the table, explain:

i) How glucose enters the blood. (3)

....

....

....

....

....

....

ii) How sodium ions enter the blood. (3)

....

....

....

....

....

....

b) Scientists concluded that glucose and sodium ions enter the epithelial cells by co-transport. Evaluate this statement. (3)

....

....

....

....

....

....

[AO2]

Q5 CFTR is a membrane transport protein which transports chloride ions from the inside of the cell to the outside down the concentration gradient followed by positively charged sodium ions, leading to a higher concentration of salts in the mucus covering the cell. In patients suffering from cystic fibrosis, the CFTR gene is mutated, so ions cannot be transported across the membrane, which causes the mucus to become thicker, blocking the bronchioles of the lung, leading to the symptoms of cystic fibrosis.

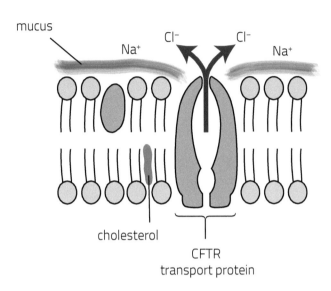

a) State the type of movement by which chloride ions cross the membrane. (1)

....

b) Explain why healthy people lack thick mucus. (3)

....

....

....

....

c) Suggest how the symptoms of cystic fibrosis could be alleviated. (2)

....

....

....

....

[AO1]

Q6 Transport by endocytosis involves the bulk movement of materials into the cell.

a) Draw a fully labelled annotated diagram to show how a bacterium is engulfed and digested by a white blood cell. (3)

b) Distinguish between the two main types of endocytosis. (4)

....

....

....

....

....

....

Question and mock answer analysis

[a & b = AO1, c = AO2]

The diagram below shows the fluid mosaic model proposed by Singer and Nicolson in 1972.

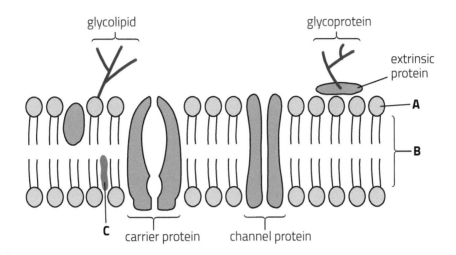

a)   State the function of C. (1)

b)   Glucose is water soluble. Vitamin A is lipid soluble. Describe and explain how the properties of the membrane shown above facilitate movement of each molecule across the membrane. (4)

c)   A student cut ten beetroot discs using a borer and then immersed them in a solution of 70% ethanol (an organic solvent) at 20°C, causing a red pigment to leak out of the cells into the ethanol turning it red. Using your knowledge of the structure of cell membranes, explain why this occurs. (2)

Isla's answers

a)   Stabilises the membrane. ✓

b)   Vitamin A
This is a non-polar molecule and so can dissolve in the phospholipid bilayer ✓ and diffuse across the bilayer. ✓

Glucose
This is a polar molecule so cannot pass through the phospholipid bilayer ✓ and instead uses the channel proteins by facilitated diffusion. ✓

> **MARKER NOTE**
> This is a particularly good answer as Isla links the property of the molecule to the method of transport well.

c)   Ethanol dissolves lipids in the bilayer ✓ so creating holes in the membrane, making it easier for the pigment to diffuse out. ✓

Isla achieves 7/7 marks

Ceri's answers

a) Makes the membrane more stable. ✓

b) Vitamin A
This is a non-polar molecule ✓ and so moves across the bilayer.

Glucose
This is a polar molecule so cannot move across the phospholipid bilayer ✓ and instead uses the channel proteins.

> **MARKER NOTE**
> Ceri has not explained the method of transport across the membrane, e.g. diffusion, facilitated diffusion.

c) Ethanol emulsifies lipids in the bilayer ✓ so creating pores for more pigment to diffuse out. ✓

Ceri achieves 5/7 marks

a) An experiment was carried out to determine how nitrate ions enter the roots of plants. The results are shown below:

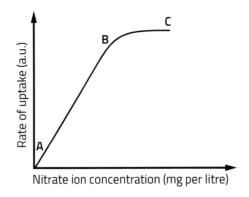

i) What conclusion can you draw about how nitrate ions are absorbed? (2)

ii) What further evidence would be needed to confirm exactly which method of uptake is involved? (2)

b) Water enters root hair cells by osmosis. The water potential (Ψ) of the soil water is $\Psi -100$kPa, and the pressure potential (ΨP) inside the root hair cell is $+200$kPa.

Calculate the solute potential (ΨS) of the root hair cell if there is no net movement of water using the formula $\Psi = \Psi S + \Psi P$. Show your working. (2)

Isla's answers

a) i) Nitrate ions use a carrier protein ✓
because above point B, further increase in ion
concentration has no effect on the rate of uptake. ✓

> MARKER NOTE
> A fuller answer would have included
> that all channel proteins are in use.

ii) I would repeat the experiment in the absence of
oxygen to see if uptake stopped. ✓
This would confirm that it is active transport
as oxygen is needed to produce ATP rather than
facilitated diffusion. ✓

> MARKER NOTE
> A perfect answer.

b) −100 −200 ✓
= −300 ✗

> MARKER NOTE
> One mark lost for omitting units.

Isla achieves 5/6 marks

Ceri's answers

a) i) Nitrate ions are absorbed by diffusion ✗
because it is directly affected by concentration
of nitrate ions between A and B. ✗

> MARKER NOTE
> Whilst it is true that rate of diffusion is
> proportional to concentration, this explanation
> fails to explain what happens at B–C.

ii) I would repeat the experiment. ✗

> MARKER NOTE
> Repeating the experiment would just provide more
> evidence and so improve reliability of the conclusion.

b) −100 −200 ✓
= −300 kPa ✓

> MARKER NOTE
> Always remember to show your full working and include units.

Ceri achieves 2/6 marks

Section 4: Enzymes and biological reactions

Topic summary

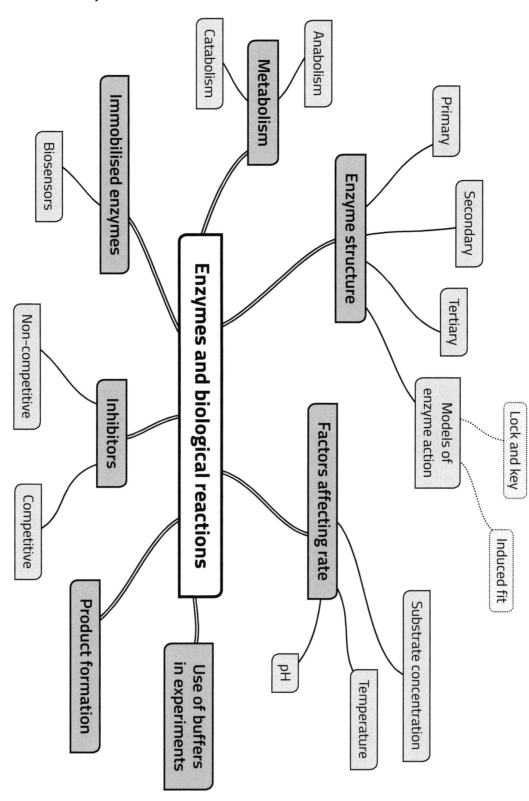

Practice questions

[AO1]

Q1 a) State what is meant by the term activation energy. (1)

....

....

b) Explain how an enzyme affects the activation energy. (2)

....

....

....

....

c) Changes in pH can cause enzymes to become inactivated or even denatured. Distinguish between the terms denatured and inactivated. (3)

....

....

....

....

....

d) In the space below, draw a graph to show the effect of pH on the rate of reaction for salivary amylase enzyme found in the mouth. (2)

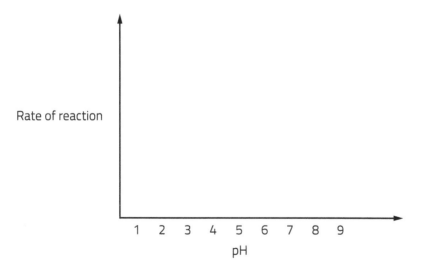

[AO2]

Q2 Methanol and ethanol are both simple organic compounds, known as alcohols, with similar chemical structures. Methanol is highly toxic and can cause blindness and even death if quantities as small as 30 cm^3 are ingested. This is because the enzyme alcohol dehydrogenase converts methanol to formaldehyde in the liver. Formaldehyde is then converted to formate by aldehyde dehydrogenase. Formate inhibits the enzymes involved in cellular respiration.

$$
\begin{array}{cc}
\begin{array}{c}
\quad\;\; H \\
\quad\;\; | \\
H - C - OH \\
\quad\;\; | \\
\quad\;\; H
\end{array}
&
\begin{array}{c}
\;\; H \quad H \qquad H \\
\;\; | \qquad | \qquad / \\
H - C - C - O \\
\;\; | \qquad | \\
\;\; H \quad H
\end{array}
\\
\text{Methanol} & \text{Ethanol}
\end{array}
$$

a) One treatment for methanol poisoning is to give the patient ethanol to drink. Using the information provided and your knowledge of enzyme kinetics, suggest why this helps prevent death. (5)

....

....

....

....

....

....

....

b) Cyanide is also a direct inhibitor of respiration. Explain how the action of cyanide differs from that of methanol. (3)

....

....

....

....

....

[AO1] [AO2]

Q3

Biosensors can be used to monitor medical conditions such as diabetes.

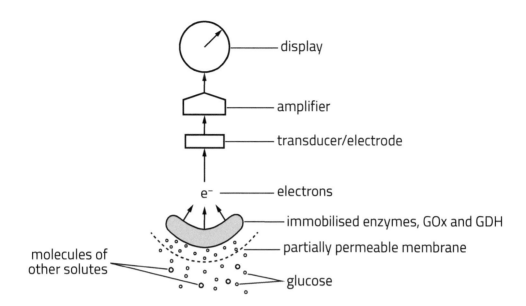

Using your own knowledge and the information included in the diagram above, describe what is meant by a biosensor and explain the advantages of using immobilised enzymes in this way. (9 QER)

[a, b = AO1] [c, d = AO2]

Q4 Cats do not produce the enzyme lactase, so are unable to digest the lactose present in cow's milk. A student carried out an experiment to produce lactose-free milk by passing milk through a column containing alginate beads with lactase enzyme immobilised inside. This is shown below:

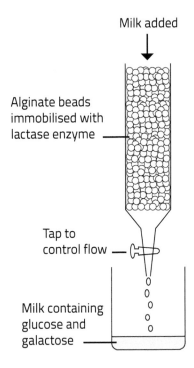

Milk added

Alginate beads immobilised with lactase enzyme

Tap to control flow

Milk containing glucose and galactose

a) Describe **two** advantages of using immobilised lactase in this way. (2)

....

....

....

....

b) Name the type of chemical reaction involved in converting lactose to glucose and galactose. (1)

....

c) The student could increase the removal of lactose from the milk by altering the flow rate. The results are shown in the graph below:

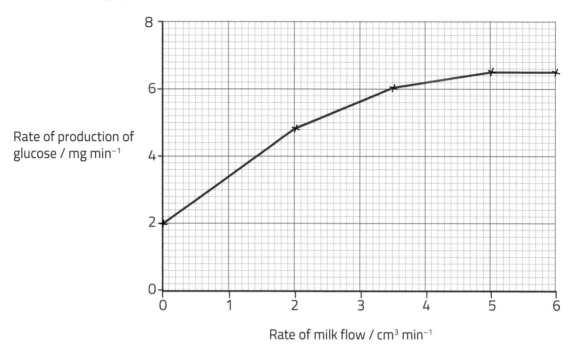

Rate of production of glucose / mg min⁻¹

Rate of milk flow / cm³ min⁻¹

EXAM TIP

Always quote data and explain all bits of the graph.

Explain the results shown. (4)

....

....

....

....

....

....

....

d) Outline **two** further ways in which the rate of lactose breakdown could be increased further. (3)

....

....

....

....

....

....

[AO2 / AO3]

Q5 Biological washing powders contain enzymes to help remove common food stains at lower temperatures. The graph below shows the activity of three enzymes commonly used in biological washing powders:

Graph showing protease activity

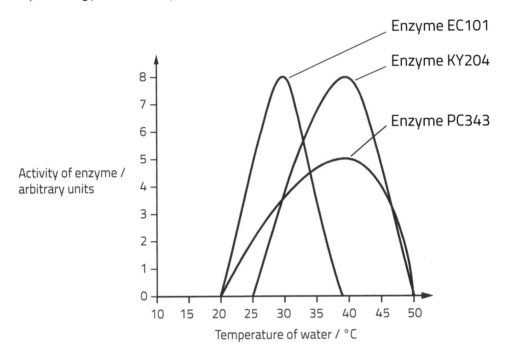

Table showing type of food stains removed by each enzyme

Protease	Common protein-based stains removed		
	Egg	Milk	Blood
Enzyme EC101	No	No	Yes
Enzyme KY204	No	Yes	No
Enzyme PC343	Yes	No	No

Using your knowledge and the data provided, answer the following questions:

a) What is the optimum temperature for your washing machine to remove most protein stains with this washing powder? (1)

....

b) Explain fully why it is not recommended to use this washing powder to remove protein-based stains at temperatures above 60°C. (4)

....

....

....

....

....

....

c) Suggest why three different enzymes are needed to remove egg, blood and milk stains. (2)

....

....

....

....

d) Explain why blood stains won't be removed when washing at 40°C. (2)

....

....

....

....

Question and mock answer analysis

[P] [AO2/AO3]

Pectin is a structural polysaccharide found in the cell walls of plant cells and in the middle lamella between cells, where it helps to bind cells together. Pectinases are enzymes that are routinely used in industry to increase the volume and clarity of fruit juice extracted from apples. The enzyme is immobilised onto the surface of a gel membrane, which is then placed inside a column. Apple pulp is added at the top, and juice is collected at the bottom.

The process is shown in the diagram below:

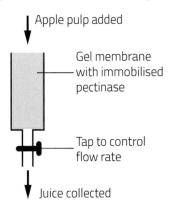

a) Explain why reducing the flow rate of material through the column would result in increased juice collected. (1)

b) The extraction of juice using pectinase was compared using equal volumes and concentrations of free enzyme, enzymes bound to the surface of a gel membrane and enzymes encapsulated inside alginate beads. The results are shown in the graph below:

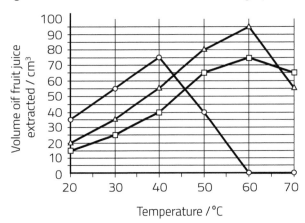

Using the graph and your own knowledge of enzymes, answer the following questions:

i) Describe and explain the results for the free enzymes at temperatures above 40°C. (4)

ii) Explain why a higher yield of juice was obtained when using free enzymes between temperatures of 20°C and 40°C than when using either kind of immobilised enzyme. (2)

iii) Explain the differences seen in the results for the enzymes bound to the gel membrane surface with those immobilised inside the beads, between temperatures of 20°C and 60°C. (2)

c) Name two variables other than flow rate that must be controlled in this experiment and explain the effect of not controlling them. (3)

Isla's answers

a) Allows more time for the pectinase to break down the apple into juice and so more enzyme–substrate complexes are formed. ✓

b) i) Above 40°C less juice is extracted, ✓ above 60°C no juice is extracted ✓ because at 60°C the enzymes are fully denatured ✓ due to the hydrogen bonds breaking.

> **MARKER NOTE**
> Isla could have explained the effect upon the active site of the hydrogen bonds breaking, i.e. that the tertiary structure deforms.

ii) The free enzymes can move about and so have more kinetic energy ✓ and so are more likely to collide with the pectinase to form ES complexes. ✗

> **MARKER NOTE**
> Isla uses the abbreviation ES – this is not an accepted abbreviation for enzyme–substrate.

iii) More juice is collected when membrane-bound enzymes are used because they directly touch the fruit. ✓ ✓

> **MARKER NOTE**
> Isla needed to explain why less juice would be extracted from enzymes immobilised inside the beads, i.e. that the substrate has to diffuse into the bead.

c) You should use apples of the same age.

> **MARKER NOTE**
> Isla should include a reason why this would affect the yield; for example, riper apples may yield more juice as cell walls begin to break down naturally with age.

The pH of apples should also be kept constant. ✓

> **MARKER NOTE**
> Isla gains one mark for two correct controlled variables and explains why it is important to control pH. Another variable to control here would be concentration of enzyme.

If the pulp was too acidic then the pectinase enzymes would become inactivated or denatured which would reduce the yield of fruit juice. ✓

Isla achieves 9/12 marks

Ceri's answers

a) There is more time for the pectinase to digest the apple. ✗

MARKER NOTE
Need to explain fully why more juice is extracted, including reference to enzyme–substrate complexes.

b) i) Above 40°C less juice is extracted, ✓ because the enzymes are denatured ✗, because the peptide bonds break. ✗

MARKER NOTE
It is not correct to say at 50°C that the enzymes are denatured as some juice is still being extracted. Many are, but not all, so better to say above 40°C enzymes are denaturing. Ceri incorrectly says that peptide bonds break.

ii) The free enzymes can move about and therefore have more kinetic energy ✓ so there are more collisions between the enzyme and substrate. ✗

MARKER NOTE
Ceri makes reference to more collisions but these must be successful, i.e. enzyme–substrate complexes form.

iii) More juice is collected from the membrane-bound enzymes. ✗

MARKER NOTE
Ceri only describes. There is no explanation, so no marks can be awarded.

c) Temperature should be kept constant, because at higher temperatures the high kinetic energy denatures the enzyme's active site. ✗

MARKER NOTE
Temperature cannot be controlled in this experiment because it is an independent variable. The consequence on the active site of high temperature is explained, but not on the yield of juice, which was the dependent variable in the experiment.

The pH should also be kept at the optimum for the enzyme say at pH 7.

MARKER NOTE
Two correct controlled variables are needed for one mark, so as only one is correct, no mark is awarded. Ceri needs to explain the effect of not controlling pH on the yield of juice.

Ceri achieves 2/12 marks

EXAMINER COMMENTARY

Be careful with abbreviations – don't use ES complexes for enzyme–substrate complexes unless you have written it out in full first. It is vital that you read the question and follow the command words to ensure that you give the answer required and so pick up full marks. When you are required to explain in terms of enzyme kinetics, make sure you include reference to enzyme–substrate complexes.

Section 5: Nucleic acids and their functions

Topic summary

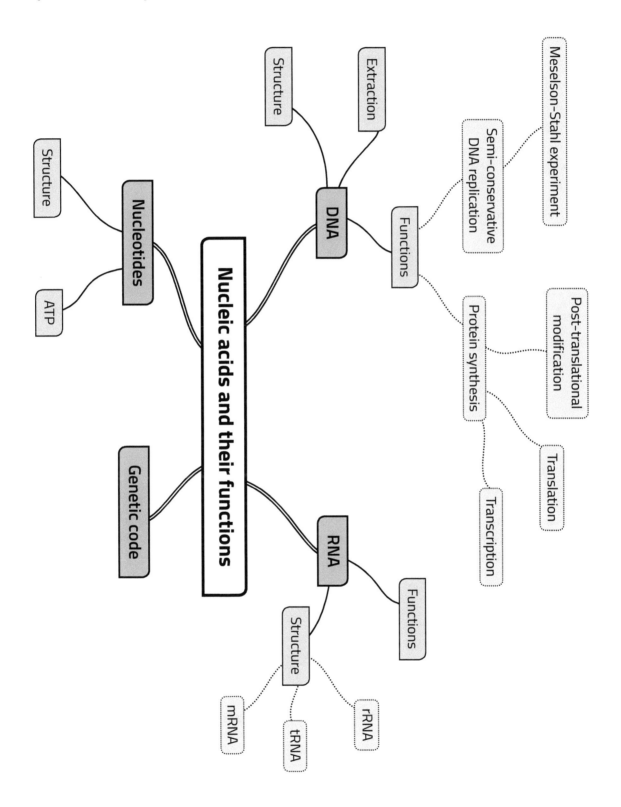

Practice questions

[M] [AO2]

Q1 Experiments by Erwin Chargaff had identified that there were four different bases in the nucleic acid found in the nucleus of cells. The table below shows the results of his experimental work:

Organism	Base composition (%)			
	A	T	G	C
Yeast	31.3	32.9	18.7	17.1
Sea urchin	32.8	32.1	17.7	18.4
Herring	27.8	27.5	22.2	22.6
Rat	28.6	28.4	21.4	21.5
Human	30.9	29.4	19.9	19.8

Source: E. Chargaff and J. Davidson, eds., *The Nucleic Acids* (New York: Academic Press, 1955)

a) Explain how the data supports the concept of complementary base pairing. (2)

....

....

....

b) Chicken anaemia virus (CAV) is a single-stranded DNA virus. Explain how the results from CAV would differ from those shown above. (2)

....

....

....

c) Watson and Crick proposed the double helix structure in 1953. If the double helix takes 3.4 nm to make one complete turn and the base pairs are 0.34 nm apart, how many base pairs would you expect in three complete turns of the helix? Show your working. (2)

Answer

....

[AO1]

Q2 The diagram below shows a component of DNA:

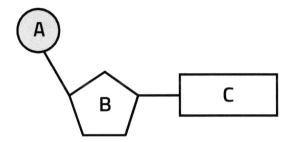

a) Name the parts A, B and C. (2)

A ....

B ....

C ....

b) Describe how a polymer of RNA would be different from a polymer of DNA. (2)

....

....

....

c) In the space below draw a labelled diagram to show the structure of tRNA. (3)

Q3

[P] [M] [AO2/AO3]

Meselson and Stahl conducted an experiment to determine the exact mechanism for DNA replication. Bacteria were grown using a heavy isotope of nitrogen ^{15}N so all DNA produced would be of a heavier weight than normal. Bacteria were then grown on a ^{14}N medium (normal weight nitrogen), and after one generation the DNA extracted formed an intermediate band halfway up the tube. The results are shown below:

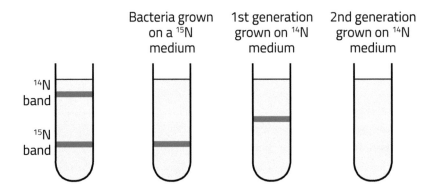

a) On the diagram above draw the results you would expect to see after growing the bacteria for a further generation on N14 medium. (1)

b) Meselson and Stahl concluded that the bacteria replicated by semi-conservative replication. How do the results from the first generation support this conclusion? (2)

c) Explain how your results for the second generation rule out dispersive replication. (2)

d) What would you expect to see if the experiment was continued for a further two generations using N14 medium? (1)

[M] [AO3]

Q4 Experiments based on work by Erwin Chargaff calculated the proportion of adenine, thymine and guanine in three different animal species containing double-stranded DNA as their nuclear material. The results are shown below:

Organism	Base composition (%)			
	A	T	G	C
Yeast	31.3			
Rat		28.4		
Human	30.9		19.9	

a) Complete the table above by estimating the proportion of other nucleotide bases in the samples. (3)

b) Explain why it would not be possible to estimate the proportion of missing nucleotides if the organisms contained single-stranded DNA. (1)

....

....

c) Scientists have suggested that bacteria living in hot springs have a much higher proportion of guanine and cytosine than bacteria living in more temperate environments. Using your knowledge of DNA structure, suggest a reason for this observation. (2)

....

....

....

....

[AO1] [AO2]

Q5

a) In eukaryotic cells, protein synthesis involves transcription and translation. State precisely in the cell where they take place. (2)

Transcription

....

Translation

....

b) In the space below, write the mRNA sequence produced during transcription from the following DNA sequence: (2)

DNA   5' AAA AGA TGA GCA TCA CCT CTT 3'

mRNA   ....

c) Use the mRNA codon table below to work out the primary sequence of the protein coded by the mRNA sequence you worked out in part b). (3)

....

....

....

		Second base				
		U	**C**	**A**	**G**	
First base	**U**	UUU ⎤ UUC ⎦ Phenylalanine UUA ⎤ UUG ⎦ Leucine	UCU ⎤ UCC ⎥ UCA ⎥ Serine UCG ⎦	UAU ⎤ UAC ⎦ Tyrosine UAA Stop codon UAG Stop codon	UGU ⎤ UGC ⎦ Cysteine UGA Stop codon UGG Tryptophan	U C A G
	C	CUU ⎤ CUC ⎥ CUA ⎥ Leucine CUG ⎦	CCU ⎤ CCC ⎥ CCA ⎥ Proline CCG ⎦	CAU ⎤ CAC ⎦ Histidine CAA ⎤ CAG ⎦ Glutamine	CGU ⎤ CGC ⎥ CGA ⎥ Arginine CGG ⎦	U C A G
	A	AUU ⎤ AUC ⎥ Isoleucine AUA ⎦ AUG Methionine start codon	ACU ⎤ ACC ⎥ ACA ⎥ Threonine ACG ⎦	AAU ⎤ AAC ⎦ Asparagine AAA ⎤ AAG ⎦ Lysine	AGU ⎤ AGC ⎦ Serine AGA ⎤ AGG ⎦ Arginine	U C A G
	G	GUU ⎤ GUC ⎥ GUA ⎥ Valine GUG ⎦	GCU ⎤ GCC ⎥ GCA ⎥ Alanine GCG ⎦	GAU ⎤ GAC ⎦ Aspartic acid GAA ⎤ GAG ⎦ Glutamic acid	GGU ⎤ GGC ⎥ GGA ⎥ Glycine GGG ⎦	U C A G

Third base

Q6

[AO1]

Describe the steps involved in synthesising haemoglobin in eukaryotic cells. (9 QER)

....

....

....

....

....

....

....

....

....

....

....

....

....

....

....

....

....

....

....

....

....

....

[AO1]

Q7 ATP is a nucleotide.

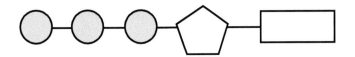

a) Label the diagram above to show the constituent molecules. (2)

b) Explain how ATP releases energy to the cell. (2)

c) Describe *two* roles of ATP in cells. (2)

d) Outline three advantages of ATP. (3)

Question and mock answer analysis

Q&A 1

[M] [AO1/AO3]

The diagram below shows a component of RNA.

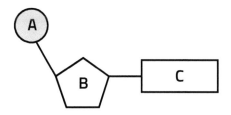

a)   Name the molecules A, B and C. (2)

b)   Describe two differences between tRNA and mRNA. (2)

c)   The table below shows the bases guanine and cytosine as percentages of the total nucleotides present in three different micro-organisms that were calculated by sequencing the genome of each organism. The scientists concluded that the virus contained single-stranded RNA, whilst the yeast contained double-stranded DNA.

Micro-organism	Base composition (%)	
	G	C
yeast	18.7	17.1
bacteria	36.0	35.7
virus	42.0	13.9

What evidence is there to support this conclusion? (3)

d)   Calculate the proportion of adenine and thymine present if the yeast contained double-stranded DNA. Show your working. (2)

Isla's answers

a) A = phosphate ✓
   B = deoxyribose ✓
   C = nitrogenous base ✓
   all correct = 2 marks.

tRNA is a clover leaf shape whereas mRNA is linear ✓
   mRNA is single-stranded whereas tRNA has double-stranded sections. ✓

b) The bacteria had very similar proportions of guanine (36.0%) and cytosine (35.7%) suggesting complementary base pairing between strands. ✓

   In the virus 42% is guanine and 13.9% is cytosine so molecule can't be double-stranded, it must be single-stranded. ✓

MARKER NOTE
To gain full marks, Isla should explain that there is no evidence to support that the molecule is RNA, as there is no reference to the proportion of thymine (which would be nil in RNA).

c) 18.7% is guanine and 17.1% is cytosine ✓
   Total = 35.8%
   Remainder = 64.2% which should be equally split between adenine and thymine so %T = 32.1% and %A = 32.1%. ✓

Isla achieves 8/9 marks

Ceri's answers

a) A = phosphate ✓
   B = pentose ✗
   C = nitrogenous base ✓
   two correct = 1 mark.

MARKER NOTE
Pentose is the type of sugar **not** the name.

b) tRNA is a clover leaf shape whereas mRNA is not ✗
   mRNA is single-stranded whereas tRNA is double-stranded. ✓

MARKER NOTE
This is not a comparison – mRNA is single-stranded.

c) There are no results to show that RNA is present, e.g. presence of uracil. ✓ The proportion of guanine and cytosine are not the same so molecule can't be double-stranded, it must be single-stranded.

MARKER NOTE
Need to say why, i.e. proportion of complementary bases the same and use data.

d) 18.7% is guanine and 17.1% is cytosine ✓
   %T = 32.1% and %A = 32.1%. ✓

MARKER NOTE
Only some of the working is shown.

Ceri achieves 5/9 marks

EXAM TIP
It is important when asked to justify or provide evidence for a conclusion that you evaluate any data provided. You should use this to support your answer.

Section 6: Cell cycle and cell division

Topic summary

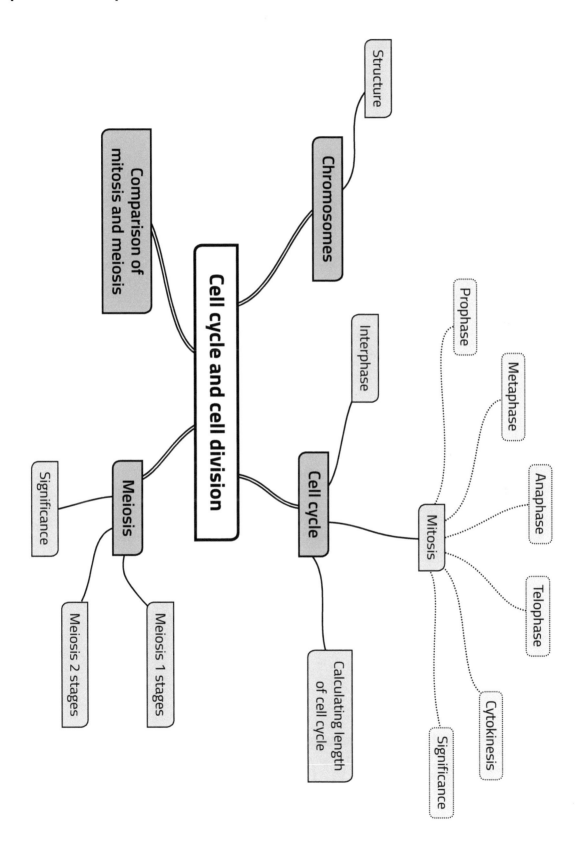

Practice questions

Q1

[AO1] [AO2]

a) Explain fully what is meant by the term allele. (2)

....

....

....

....

b) In the space below, draw a labelled diagram to show a chromosome following DNA replication. (2)

c) *Aedes agypti* is a species of mosquito which belongs to the family Culicidae where all members have a diploid chromosome number 2n = 6 except *Chagasia bathana* which has a diploid number of 2n = 8. The following diagrams show chromosomes undergoing mitosis and meiosis:

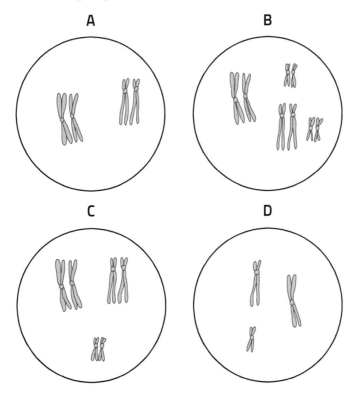

A

B

C

D

i) Explain what is meant by diploid. (1)

....

....

ii) Identify which cells belong to the mosquito *Aedes agypti*, and which to *Chagasia bathana*. (3)

Aedes agypti ....

Chagasia bathana ....

iii) In the space below draw the haploid set of chromosomes you would expect to see from *Chagasia bathana*. (2)

[P] [M] [AO2] [AO3]

Q2 A student performed a root tip squash with the roots from three onion plants. The tips were stained and viewed under a microscope. The appearance of the chromosomes is shown below:

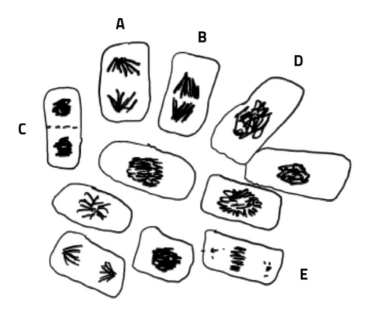

a) Identify which stages of mitosis are represented by drawings A–E. (3)

A ....

B ....

C ....

D ....

E ....

b) A and B are both the same stage. Suggest why A and B appear different. (2)

....

....

....

....

c) The student observed 120 cells and recorded the numbers in each sample that were in prophase, metaphase, anaphase or telophase. The results are shown below:

Root tip sample	Number of cells seen in				% cells undergoing mitosis
	Prophase	Metaphase	Anaphase	Telophase	
1	12	5	3	7	
2	21	6	2	8	
3	18	3	2	9	
Mean					

i) Complete the table to show the mean number of cells observed (to 1 decimal place).

ii) The number of cells observed in each phase is directly proportional to the length of that phase. Using the mean data, put the stages into order with the shortest stage first, and the longest stage last. (2)

....

....

....

....

iii) Calculate the percentage of cells undergoing mitosis in each sample. (2)

Answer

....

iv) The student repeated the experiment taking samples from the root cap itself, and 0.5 mm, 1.0 mm and 1.5 mm behind the root cap. They found that the mitotic index (percentage of cells undergoing mitosis) was highest in cells taken from the root cap and 0.5 mm from the cap, and then decreased rapidly further away from the cap. Suggest a reason for the results seen. (2)

....

....

v) If the length of the cell cycle in these cells is 22 hours, calculate the mean length of prophase to the nearest whole minute. Show your working. (3)

.... min

[a = AO1] [b, c = AO2]

Q3

a) Explain what is meant by the following terms. (5)

i) Homologous

....

....

ii) Polyploidy

....

....

iii) Cytokinesis

....

....

iv) Allele

....

....

v) Interphase

....

....

b) Describe how mitosis differs from meiosis. (4)

....

....

....

....

....

....

c) The following diagram shows how the mass of DNA changes during meiosis: (2)

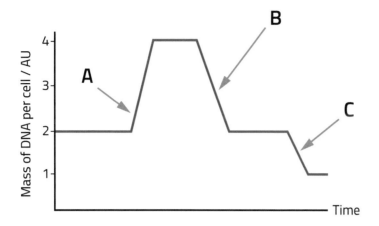

i) Explain why the graph shows meiosis rather than mitosis. (2)

....

....

....

....

ii) Explain how the DNA content is halved at point B. (2)

....

....

....

....

iii) Explain why during process C the DNA content is halved but the chromosome number remains the same. (3)

....

....

....

....

....

Q4

[AO1]

Explain, with the aid of annotated diagrams, the significance of meiosis and how it is achieved through cell division. (9 QER)

....

....

....

....

....

....

....

....

....

....

....

....

....

....

....

....

....

....

....

....

....

....

....

Question and mock answer analysis

[P] [a,b = AO1, c) = AO2/AO3]

The diagram below shows the relative lengths of the cell cycle in actively dividing cells taken from the root tip of a garlic plant:

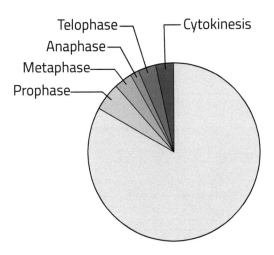

a) Describe the changes that occur to the nucleus of a plant cell during anaphase. (3)

b) Explain how cytokinesis differs in plant cells from that in animal cells. (3)

c) A new drug has been developed that inhibits mitosis by preventing the formation of the spindle fibres. Garlic bulbs were grown in a solution of the new drug and the quantity of DNA present in a cell from the root tip was measured over the 24-hour length of the cell cycle. The results are shown below together with the results from garlic bulbs grown in water.

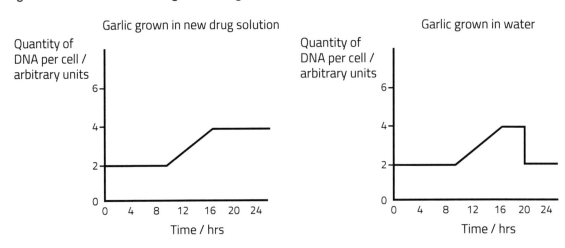

i) Using your knowledge of the cell cycle, explain how the results of this experiment show that the new drug inhibits mitosis. (3)

ii) Explain how you could verify this experimentally. (3)

Isla's answers

a) During anaphase the centromeres divide ✓ and the spindle fibres shorten ✓ pulling the chromatids towards opposite poles of the cell. ✓

b) In plant cells the cellulose cell wall prevents the cell from cleaving in two ✓ so instead a cell plate develops from the centre outwards dividing the cell in two. ✓

MARKER NOTE
Isla did not say how cytokinesis occurs in animal cells, i.e. via the membrane infolding.

c) i) Interphase occurs in both experiments as DNA quantity doubles from 2 to 4 arbitrary units between 10 and 16 hours. ✓

MARKER NOTE
Good to reference data.

When grown in the new drug, there is no halving of DNA at 20 hours as seen in the garlic grown in water. ✓ This shows that there is no cytokinesis because the lack of spindle fibres prevents anaphase. ✓

ii) You could take a sample of garlic root from both plants and stain with a suitable stain to show chromosomes, e.g. acetic orcein ✓ and then squash sample between two glass slides. When viewed under the microscope cells grown in water should show all parts of the cell cycle, e.g. prophase, metaphase, anaphase, telophase and cytokinesis. ✓ Cells grown in the new drug solution would not show spindle fibre formation in anaphase or any cells in telophase. ✓

Isla achieves 11/12 marks

Ceri's answers

a) During anaphase the spindle fibres shorten ✓ pulling the chromatids towards opposite ends of the cell. ✓

MARKER NOTE
Ceri did not mention about the centromeres dividing.

b) In animal cells the membrane infolds, dividing the cell into two ✓ whereas in plant cells the cellulose cell wall prevents the cell from dividing into two. ✓

MARKER NOTE
Ceri did not mention about the growth of a cell plate.

c) i) When placed in the drug, DNA does not decrease, ✗ so cell division has not happened. ✗

MARKER NOTE
Reference to DNA decreasing is too vague, and there is no reference to data either DNA amount or time. The question asks how the results show that the new drug inhibits mitosis: reference to cell division not occurring is also too vague, when information in the question details that the new drug prevents spindle formation. Ceri should link this to anaphase not occurring, and that subsequently cytokinesis will not occur.

ii) You could take a sample of garlic root from both plants and view under the microscope. Cells grown in the new drug solution would not show spindle fibre formation. ✓

MARKER NOTE
Ceri now makes reference to how this would be done, e.g. staining cells or performing root tip squash. Though Ceri does mention that no spindle fibres would be seen, we would expect details of no cells in telophase to be included as this demonstrates knowledge of the stages in the cell cycle.

Ceri achieves 5/12 marks

EXAM TIP
Learn the key events in mitosis (and meiosis) and be able to recognise drawings of each stage. You should be able to use data given to reach or support a conclusion.

Component 2: Biodiversity and Physiology of Body Systems

Section 1: Classification and biodiversity

Topic summary

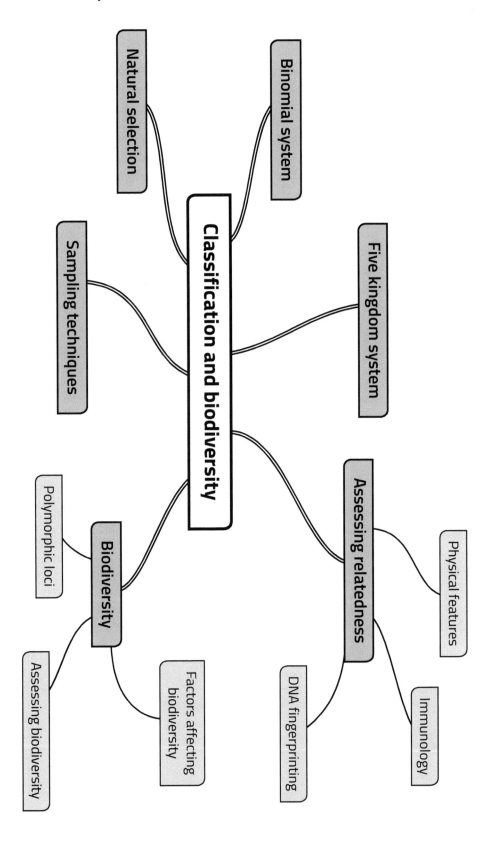

Practice questions

Q1

[P] [M] [AO1] [AO2]

One method for estimating the size of a terrestrial animal population is mark-release-recapture. A student using a pitfall trap caught 21 beetles in a 24-hour period. The beetles were carefully marked and released; 24 hours later the trap was reset and the number caught in the second sample was 18, of which 9 were marked.

The population size can be estimated by using the following equation:

Total number of animals in the population $= \dfrac{N_1 \times N_2}{N_m}$

Where N_1 = number in sample 1

N_2 = number in sample 2

N_m = number marked in sample 2

a) i)   Estimate total number of animals in the population. Show your working. (2)

Answer = ...

ii) Explain why sampling should be conducted at random. (1)

....

....

iii) Explain what precautions must be observed during the collection process. (2)

....

....

....

....

iv) Describe one assumption that has to be made when using the results. (1)

....

....

[M] [P] [AO2] [AO3]

Q2 A student wanted to assess the effect of farming on biodiversity. $1\,m^2$ quadrats were placed in a field which had been intensively farmed but had been left to lie fallow for 12 months (Field A), i.e. had not been reseeded, grazed or managed for 12 months, and the number of different species and number of each species were recorded in 10 quadrats. The experiment was repeated in a field which had not been farmed, seeded or grazed for 5 years (Field B). The results are shown below. N is the total number of plants in a field.

a) Suggest two variables which should be controlled during the experiment. (2)

....

....

....

b) Calculate the Simpson diversity index for both fields by completing the table and using the formula below: (6)

Species	Field A Number of plants per m^2 in recently farmed field (n_1)	$n_1(n_1-1)$	Field B Number of plants per m^2 in unfarmed field (n_2)	$n_2(n_2-1)$
Buttercup	1		12	
Daisy	3		8	
Plantain	0		9	
Clover	0		13	
Thistle	0		12	
Dandelion	1		11	
Nettle	1		0	
	N = N−1=	$\Sigma n_1(n_1-1) =$	N = N−1=	$\Sigma n_2(n_2-1) =$

$$S = 1 - \frac{\Sigma n(n-1)}{N(N-1)}$$

Field A

S = ...

Field B

S = ...

c) The student concluded that field B had the highest diversity. Explain how confidence in the conclusion could be improved. (2)

....

....

....

....

d) Suggest why farmers are encouraged to let fields lie fallow for several years. (1)

....

....

....

[AO2] [AO3]

Q3 The following drawings show limbs from four different mammals:

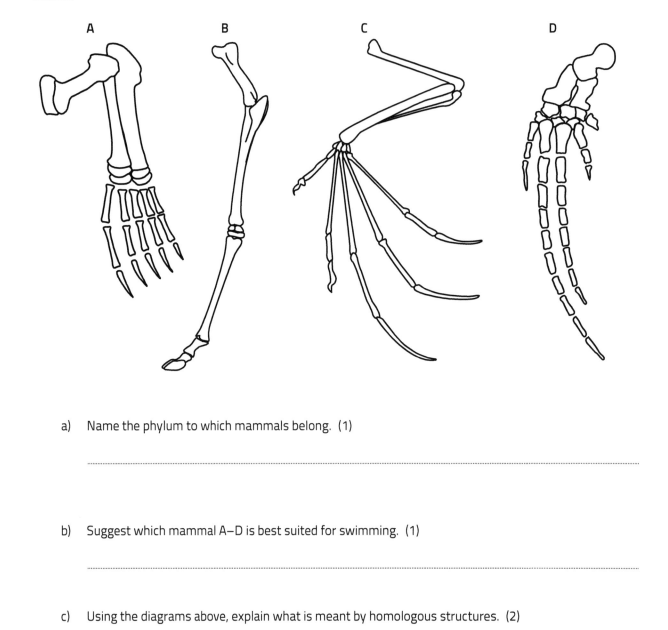

a) Name the phylum to which mammals belong. (1)

....

b) Suggest which mammal A–D is best suited for swimming. (1)

....

c) Using the diagrams above, explain what is meant by homologous structures. (2)

....

....

....

....

d) Explain why the limbs shown provide evidence for divergent evolution. (1)

....

....

e)   The following diagram shows how mammals are related to other animals:

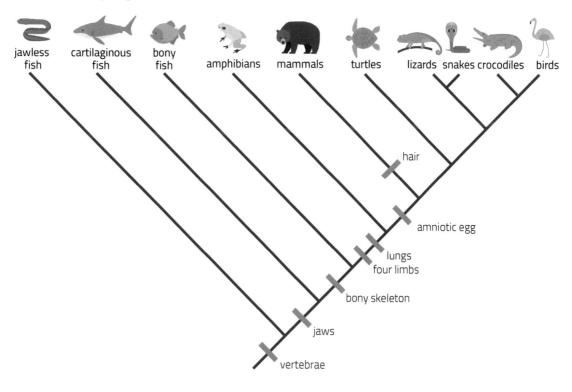

i)   Name the type of diagram shown. (1)

....

ii)   Using the diagram, identify which animals have:

Hair...(1)

Lungs ....(1)

Q4

[AO1]

Using examples, explain how natural selection has led to anatomical, physiological and behavioural adaptations in mammals. (9 QER)

Question and mock answer analysis

Q&A 1

[AO2] [AO3]

The diagrams below show skulls from three different primates, and their phylogenetic tree:

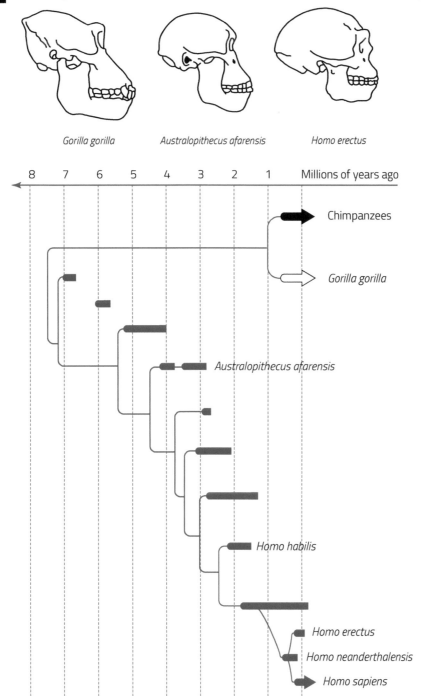

Gorilla gorilla *Australopithecus afarensis* *Homo erectus*

a) Name the class to which all these primates belong. (1)

b) Define the term species. (2)

c) i) With reference to the diagrams, suggest why scientists regard *Homo erectus* as being more closely related to *Australopithecus afarensis* than *Gorilla gorilla*. (3)

   ii) Using their classification, identify which primate is most closely related to modern humans, and explain your answer. (2)

Isla's answers

a) Vertebrata. ✗

> MARKER NOTE
> Vertebrata is the phylum, not class.

b) A group of organisms with similar characteristics that can interbreed ✓ to produce fertile offspring. ✓

c) i) The shape of the jaw and cranium of Homo erectus and Australopithecus afarensis look similar in shape. ✓ Homo erectus shares a more recent common ancestor with Australopithecus afarensis about 4–5 million years ago ✓ whilst it is 7–8 million years ago for Gorilla gorilla. ✓

   ii) Homo erectus, ✓ because erectus and sapiens share same genus. ✓

Isla achieves 7/8 marks

Ceri's answers

a) Mammals. ✓

b) A group of organisms that can breed ✗ to produce fertile offspring. ✓

> MARKER NOTE
> To be a member of the same species, organisms need to interbreed (or breed together).

c) i) The skulls look similar. ✗

> MARKER NOTE
> Similar skulls is too vague. Ceri needs to be specific about the shape of the skull, jaw, cranium or teeth, and to include which primates are being referred to.

   They share an earlier common ancestor. ✓

> MARKER NOTE
> This is correct, but more information could be used which is in the graph, i.e. Gorilla gorilla shares a common ancestor 7–8 million years ago whilst Australopithecus afarensis is only 4–5 million years ago.

   ii) Homo erectus, ✓ because erectus and sapiens share same genus. ✓

> MARKER NOTE
> Ceri should have used their classification as asked, rather than referring to the diagram.

Ceri achieves 5/8 marks

EXAM TIP
Always use all the data available from the question and tables/graphs to inform your answer.

Section 2: Adaptations for gas exchange

Topic summary

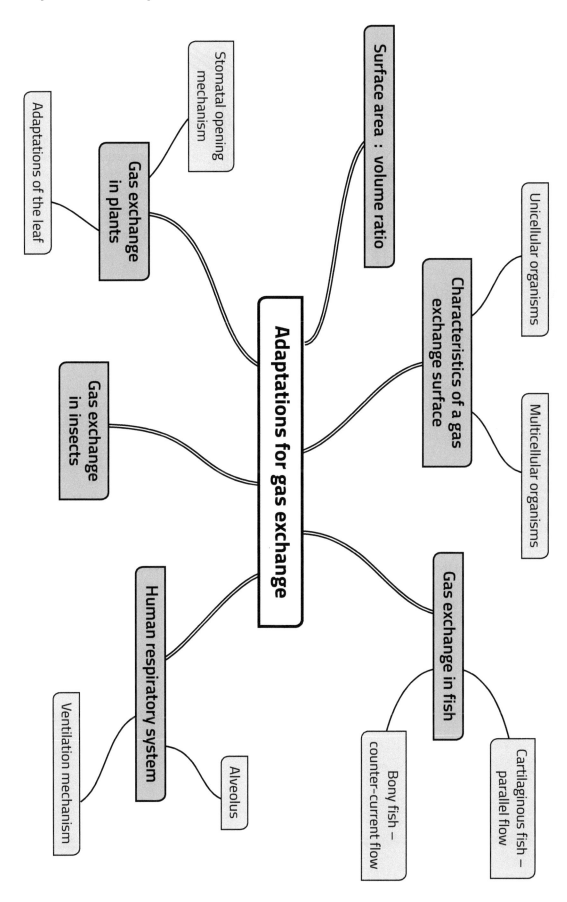

Practice questions

Q1

[AO2]

The following diagram shows one alveolus from a mammalian lung:

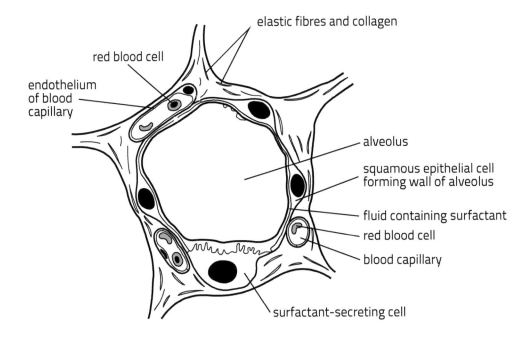

a) Explain how three features **shown** in the diagram aid gas exchange. (3)

....

....

....

....

....

b) Premature babies can develop respiratory distress syndrome (RDS) because insufficient surfactant is produced by the surfactant-secreting cells. As a result, these babies have difficulty in expanding their lungs, and absorbing sufficient oxygen. Babies are given artificial surfactant containing, e.g. dipalmitoylphosphatidylcholine (DPPC), which is a phospholipid. The diagram opposite shows the action of DPPC in premature babies.

Without DPPC

With DPPC

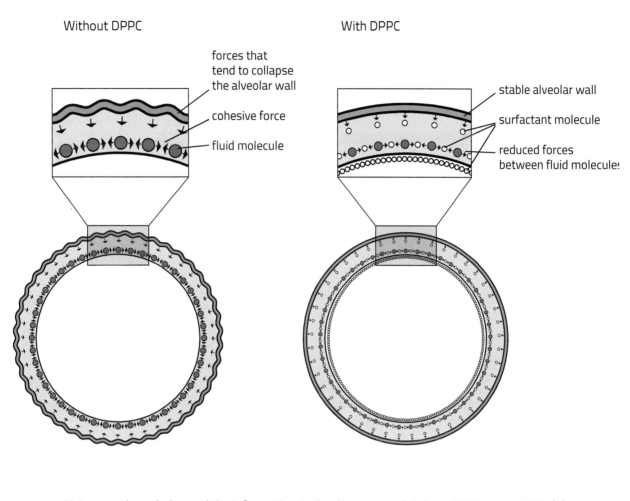

forces that
tend to collapse
the alveolar wall

cohesive force

fluid molecule

stable alveolar wall

surfactant molecule

reduced forces
between fluid molecules

Using your knowledge and the information in the diagram, explain how DPPC treats RDS. (4)

....

....

....

....

....

....

....

(AO1] [AO2]

Q2 The following diagram shows how gills function in bony fish:

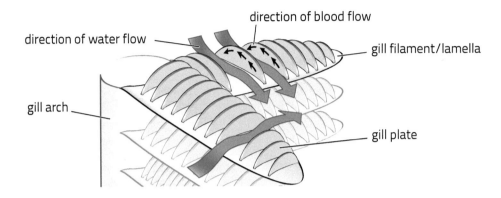

a) Name the type of flow seen in bony fish. (1)

....

....

b) The following graph shows the oxygen saturation across the length of the gill lamellae in a bony and cartilaginous fish. Using the information in the diagram, graph and your own knowledge, explain why gas exchange in bony fish is more efficient than in cartilaginous fish. (4)

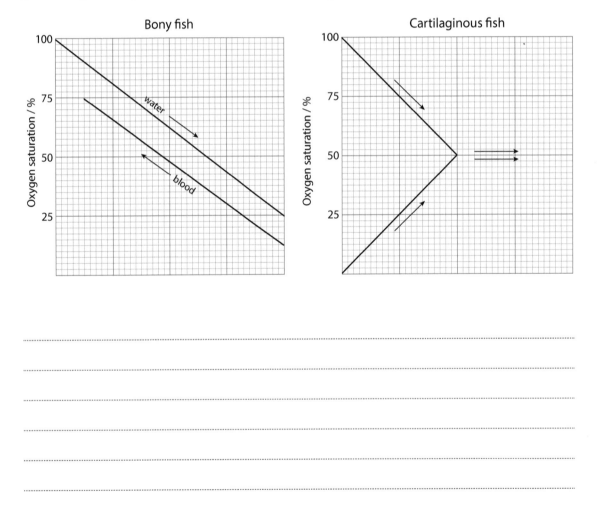

....

....

....

....

....

....

c)   Give two reasons why gills do not function effectively on land. (2)

....

....

....

....

....

Q3

[P] [AO1] [AO2] [AO3]

Gas exchange in plants is made possible due to the presence of thousands of tiny pores called stomata. The guard cells surrounding each stoma control the size of each pore to reduce water loss by transpiration.

a) Define transpiration. (1)

....

....

b) In an experiment to investigate the stomatal opening mechanism, scientists grew genetically identical plants in growth media containing different concentrations of potassium ions. They measured the concentration of malate in leaves and recorded the percentage of stomata open when exposed to a constant light source. The results are shown below:

Concentration of potassium ions / mmol	Concentration of malate ions / mmol	Percentage of stomata open / %
0.0	0.1	5
5.0	0.2	12
10.0	0.8	41
15.0	1.4	64
20.0	1.8	82
25.0	2.0	98

i) Plot a graph to show the effect of potassium ion concentration on the percentage of stomata open. (5)

ii) How does the data support the stomatal opening mechanism theory? (4)

....

....

....

....

....

....

iii) Outline what additional variables would need to be controlled to increase the confidence of the results seen. (2)

....

....

....

....

Q4

[AO1]

All gas exchange surfaces share several common features, including a large surface area to volume ratio.

a) Explain how three *other* features present in all living organisms increase the exchange of gases required for respiration. (3)

....

....

....

....

....

....

b) Explain how one feature not present in single-celled organisms increases the exchange of gases required for respiration. (1)

....

....

Q5

[AO2] [AO3]

Explain how different terrestrial animals have become adapted for gas exchange in their individual environments. (9 QER)

Question and mock answer analysis

[AO2] [AO3]

Q&A 1 The following diagram shows the tracheal system of an insect. Using the diagram and your knowledge answer the questions that follow.

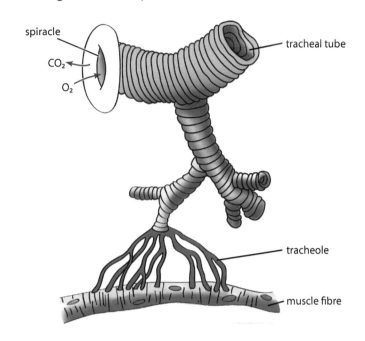

a)   List three adaptations for gas exchange shown. (2)

b)   Describe how one adaptation listed in part a) aids gas exchange. (1)

c)   Using the diagram and your knowledge, explain why insects are adapted to dry environments. (2)

Isla's answers

a) *Tracheal system provides a large surface area, the tracheole walls are thin ✓ and it has a good blood supply. ✗*

> **MARKER NOTE**
> Only two are correctly listed so one mark awarded.

> **MARKER NOTE**
> Insects do not have a good blood supply – this is a general characteristic of many gas exchange surfaces but not insects.

b) *Tracheole ends are filled with fluid to allow gases to dissolve before diffusing into the muscle fibre. ✓*

c) *Insects have spiracles which can close ✓ and the tubes are lined with chitin, to prevent water loss. ✓*

Isla achieves 4/5 marks

Ceri's answers

a) The cell walls are thin ✗
   and moist. ✗

> **MARKER NOTE**
> Ceri has used the term cell wall rather than tracheole wall which is incorrect as animal cells don't have cell walls.

> **MARKER NOTE**
> The ends of the tracheoles are fluid filled, the tracheole walls are not moist.

b) Thin cell walls means that
   diffusion occurs faster. ✗

> **MARKER NOTE**
> Ceri hasn't been penalised again for the use of cell walls but would need to say that the diffusion distance is reduced NOT that diffusion occurs faster.

c) The spiracles can close to
   prevent water loss. ✓

> **MARKER NOTE**
> Ceri fails to list a second adaptation, e.g. that the tubes are lined with chitin, which reduces water loss because it is waterproof.

Ceri achieves 1/5 marks

EXAM TIP
Remember you may have to give three points for just two marks – read the question carefully. Whilst there are many shared features of gas exchange surfaces, some do not apply to all.

[AO2] [AO3]

Q&A 2 Students were asked to set up an experiment to investigate water loss by a plant. The instructions are given below:

A. Cut two leafy shoots.

B. Cover the leaves of one shoot with petroleum jelly (Vaseline).

C. Place the shoots in separate beakers of water and cover surface of water with oil.

D. Record the total mass of each experimental setup.

E. Expose the shoots to light and weigh them again at 30-minute intervals for 5 hours. Then calculate the percentage change in mass.

The students observed that the percentage change in mass of the shoot with petroleum jelly on its leaves was less than the shoot with no petroleum jelly. The students concluded that the percentage change in mass of water lost from the shoot was equal to the mass of water absorbed by the shoot.

a) Explain why the students would be incorrect in reaching this conclusion. (3)

b) Explain how the leaf is adapted for gas exchange. (3)

Component 2 Practice questions

Isla's answers

a) *Some of the water taken into the plant would be needed for other things such as photosynthesis* ✓ *and maintaining turgidity.* ✓

MARKER NOTE

Isla could have included that some water is produced in respiration.

b) *Leaves are thin and flat providing a large surface area. They have many pores called stomata and large spaces in the spongy mesophyll.* ✓ ✓

MARKER NOTE

Isla lists three adaptations so is worthy of credit but fails to explain how they adapt the leaf, i.e. large surface area for gas exchange, stomata allows gases to enter and leave the leaf, and air space allows gases to diffuse.

Isla achieves 4/6 marks

Ceri's answers

a) *Water is also needed maintaining turgidity.* ✓

MARKER NOTE

Ceri could also have included that some water is produced in respiration and needed for photosynthesis.

b) *Leaves have a large surface area for gas exchange* ✓ *by being thin and flat,* ✓ *and have many pores called stomata.*

MARKER NOTE

Ceri did not explain the role of the stomata in allowing gases into the leaf and could have included the role of the air spaces in diffusion.

Ceri achieves 3/6 marks

Section 3: Adaptations for transport

Topic summary

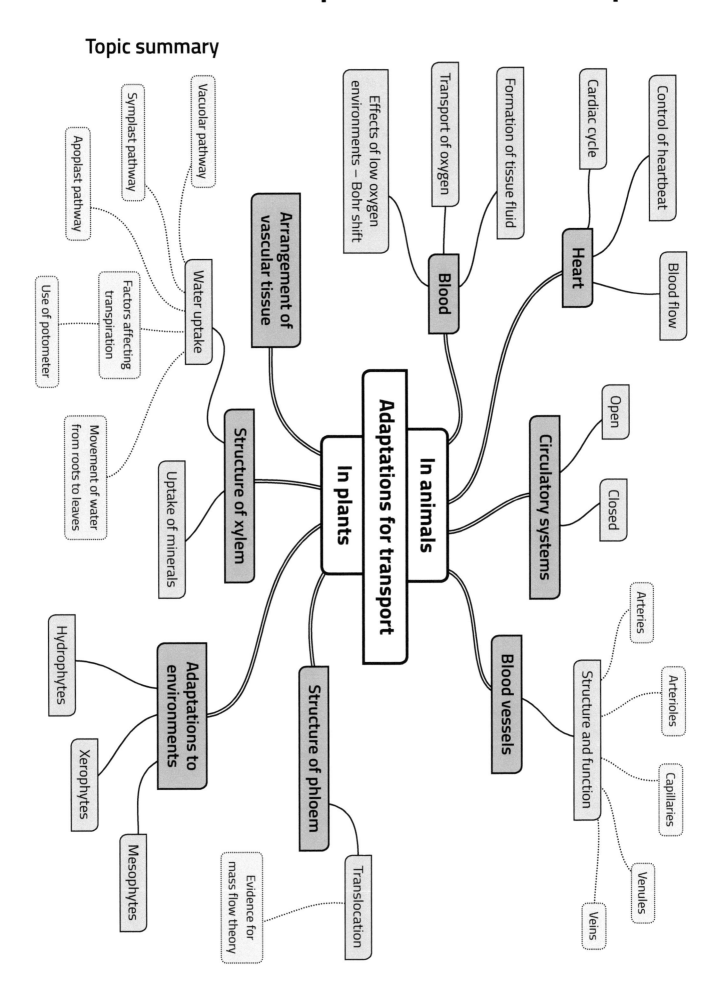

Practice questions

[AO1]

Q1 With the aid of labelled diagrams explain how the structure of arteries, capillaries and veins enables them to carry out their function. (9 QER)

....

....

....

....

....

....

....

....

[M] [AO2]

Q2

The following graph shows an electrocardiogram taken on a healthy middle-aged male:

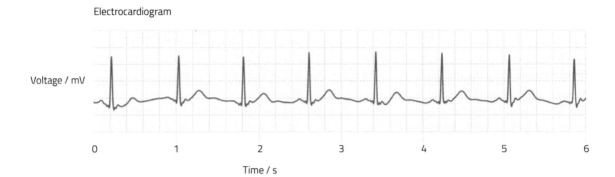

Electrocardiogram

Voltage / mV

Time / s

a) Calculate the resting heart rate for the patient to the nearest beat per minute (bpm). Show your working. (2)

Answer: ....

b) Mark the P, QRS and T events on the graph above for one cardiac cycle. (1)

c) Detail the events that occur in the heart between 1.0 and 1.1 seconds shown on the trace. (4)

....

....

....

....

....

....

d) Label the isoelectric line on the trace. Explain what is occurring during this time. (1)

....

....

Q3

[M] [AO2] [AO1]

The following graph shows the dissociation of oxyhaemoglobin at varying partial pressures of oxygen for a mammal at sea level:

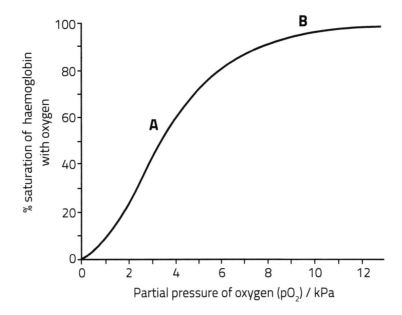

a) Using your knowledge of the structure of haemoglobin, explain the shape of the curve at part A and B and its advantage to the mammal. (4)

....

....

....

....

....

....

b) If each red blood cell contains around 280×10^6 oxyhaemoglobin molecules when 98% saturated, how many oxygen molecules are released from one red blood cell at 49% saturation? Show your working. (3)

Answer: ....

c) Foetal haemoglobin has a higher affinity for oxygen than maternal haemoglobin.

   i) On the graph. draw and label a curve to show the dissociation of foetal haemoglobin. (1)

   ii) Explain the advantage to the foetus of having an oxygen dissociation curve like the one you have drawn in part (i). (2)

   ....

   ....

   ....

   ....

   ....

d) Carbon dioxide concentration affects the dissociation of oxyhaemoglobin in mammals.

   i) Draw and label a curve on the graph to show the dissociation curve of a normal mammal under increased carbon dioxide concentrations. (1)

   ii) Explain how rising carbon dioxide concentration in the blood results in increased dissociation of oxyhaemoglobin. (5)

   ....

   ....

   ....

   ....

   ....

   ....

   ....

Q4 [AO1] [AO2]

Kwashiorkor is a severe form of malnutrition found in some developing regions where babies and young children have insufficient protein in their diets. Tissue fluid collects under the skin causing swelling called oedema.

a) Describe how tissue fluid is formed in healthy people. (4)

....

....

....

....

....

....

b) Explain why children with low protein diets develop oedema. (3)

....

....

....

....

....

[AO2]

Q5 The diagram below shows the changes in blood pressure in the left atrium, left ventricle, and aorta during one cardiac cycle. Using your knowledge and the information in the graph, answer the questions that follow.

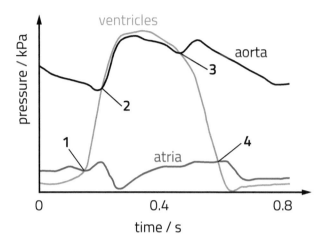

a) Which number 1, 2, 3 or 4 represents the point at which the aortic semi-lunar valve opens? Explain your answer. (2).

Answer: ....

Reason: ....

....

b) Which number 1, 2, 3 or 4 represents the point at which the left atrio-ventricular valve closes? Explain your answer. (2).

Answer: ....

Reason: ....

....

c) Why does the blood pressure in the aorta not fall to zero? (3)

....

....

....

....

[P] [AO1] [AO2]

Q6 The following photograph shows a section of a water lily as seen under the light microscope:

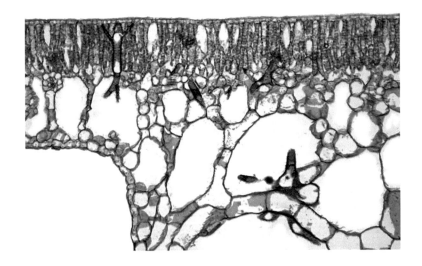

a) Name this type of plant which is adapted to living submerged in water. (1)

....

b) Label on the photograph *three* features which enable the plant to survive submerged in water. Explain how each feature accomplishes this. (6)

Feature: ....

Explanation: ....

....

....

Feature: ....

Explanation: ....

....

....

....

Feature: ....

Explanation: ....

....

....

Q7

[AO1]

Explain how the structure of the cells in the root cortex prevents and allows the different mechanisms by which water moves across the cortex from the epidermis into the xylem. (9 QER)

Q8

[P] [AO3]

An experiment was carried out to show which vessels transport solutes in plants. Carbon dioxide was radioactively labelled using ^{14}C, and a plant allowed to photosynthesise under optimal conditions for 24 hours. Sections of the plant stem were then taken and exposed to X-ray film, producing the autoradiograph shown below:

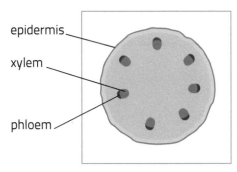

epidermis

xylem

phloem

section of stem
placed against photographic
film in the dark

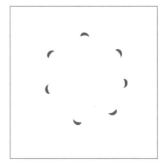

developed film emulsion
is fogged by the presence
of radioactivity in the phloem

a) What can be concluded from the autoradiograph results shown? (2)

....

....

....

....

....

b) State **two** variables that would have been controlled in order to ensure 'optimal conditions' were maintained. (2)

....

....

....

[M] [P] [AO3]

Q9 A potometer was used to measure rate of transpiration in a leafy shoot placed in different temperature environments. The apparatus was set up as shown below and repeated using the same leafy shoot. At each temperature, the plant was given 30 minutes to equilibrate before readings commenced. The diameter of the capillary tube was 1.0 mm, and volume = $\pi r^2 \times d$ (where π = 3.14, r = radius of capillary tube and d = distance bubble travels).

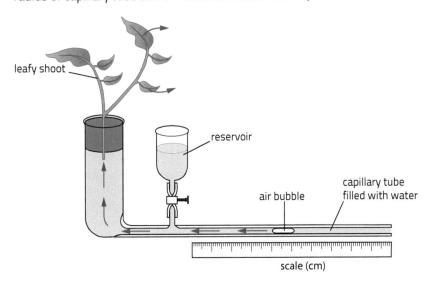

The following results were obtained:

Temperature /°C	Distance bubble moved in one minute / cm	Distance bubble moved in one minute / cm	Distance bubble moved in one minute / cm	Average distance bubble moved in one minute / cm	Average volume water lost mm³ per min
10	1.1	1.3	0.9		
20	2.3	2.1	2.3		
30	4.1	4.5	4.7		
40	8.9	8.3	8.1		
50	2.2	2.3	2.5		

a) Calculate the average distance bubble moved and volume of water lost and add to the table. In the space below show your working. (3)

Answer: ....

b)   Explain *two* precautions which should be used when setting up the apparatus. (2)

....

....

....

....

c)   Other than humidity, name *two* other variables which should be controlled during the experiment. Suggest how you would control them. (2)

....

....

....

....

d)   Draw a conclusion from the results. (3)

....

....

....

....

....

....

....

Q10 [AO1]

Distinguish between the following terms:

a) Adhesive and cohesive forces (2)

b) Transpiration and capillarity (2)

c) Source and sink (2)

d) Apoplast and symplast pathways (2)

Question and mock answer analysis

[AO1] [AO2]

With the aid of a labelled graph, describe how oxygen and carbon dioxide are carried in the blood. Suggest why oxygen is released more readily in muscles where lactic acid has built up. (9 QER)

Isla's answers

Oxygen binds to haemoglobin within red blood cells, forming oxyhaemoglobin via a reversible reaction. Each molecule of haemoglobin can carry four molecules of oxygen, one attached to each of the four haem groups. As oxygen binds, the haemoglobin molecule changes slightly, making it easier for the next one to bind. This is known as co-operative binding. A large increase in partial pressure is needed for the fourth and final oxygen molecule to bind.

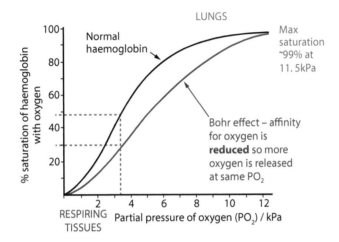

Some carbon dioxide is carried in red blood cells as carbamino-haemoglobin, but most is carried has bicarbonate ions (HCO_3^-) ✓. Carbon dioxide is converted into HCO_3^- inside the red blood cells, following a reaction involving the enzyme carbonic anhydrase. ✓

HCO_3^- diffuses out of the red cell and into the plasma where it is transported to the lungs ✓. During exercise, muscles begin to respire anaerobically and so glucose is converted to lactic acid which builds up in muscles. ✓ The increased lactic acid concentration in the muscles lowers the pH of blood as protons are released from lactic acid. This reduces haemoglobin's oxygen affinity, which is called the Bohr effect ✓, and causes oxyhaemoglobin to dissociate more easily, which is seen by a shift to the right in the oxygen dissociation curve ✓. This is an advantage during exercise, as oxygen is released more readily to respiring tissues. ✓

MARKER NOTE

Isla could have included details of the dissociation of carbonic acid to hydrogen and hydrogen carbonate ions and the chloride shift that occurs to maintain electrochemical neutrality via facilitated diffusion.

MARKER NOTE

Isla could have expanded her answer to include reference to H^+ ions binding to oxyhaemoglobin which releases oxygen

Isla achieves 7/9 marks

EXAMINER COMMENTARY

Isla constructs an articulate, integrated account, which shows sequential reasoning. The answer fully addresses the question with no irrelevant inclusions or significant omissions. The candidate uses scientific conventions, vocabulary and spelling appropriately and accurately. A good annotated graph helps to support her account.

Ceri's answers

Oxygen is carried in red blood cells as oxyhaemoglobin, with each molecule of haemoglobin carrying four molecules of oxygen.

Carbon dioxide is produced as a waste gas which is transported in the plasma to the lungs where it is eliminated. Some carbon dioxide is carried in red blood cells as carbamino-haemoglobin. ✓ When exercising, more carbon dioxide is produced which needs to be eliminated. We breathe faster and harder to help to get rid of it.

Carbon dioxide is carried in the blood dissolved in the plasma. ✓

The dissociation curve changes when there is lots of lactic acid, it shifts to the right, something called the Bohr shift. ✓

MARKER NOTE

Ceri needs to include a graph to illustrate co-operative binding of oxygen and haemoglobin.

MARKER NOTE

Ceri needs to link high lactic acid concentrations to a fall in pH.

MARKER NOTE

Ceri should expand how carbon dioxide is carried, i.e. as hydrogen carbonate ions in the blood, and some as carbamino-haemoglobin in the red blood cells.

MARKER NOTE

The Bohr shift is mentioned but Ceri needs to demonstrate understanding; for example, the effect on haemoglobin's affinity for oxygen, and how this releases oxygen more easily.

Ceri achieves 3/9 marks

EXAMINER COMMENTARY

Ceri makes some relevant points, such as those in the indicative content, but shows limited reasoning. The answer addresses the question but with significant omissions. Ceri has limited use of scientific conventions and vocabulary.

[AO1] [AO2]

Q&A 2 Scientists carried out an experiment to investigate which vessels in a plant were used to translocate solutes. They placed two identical plants in a gas jar, and added carbon dioxide containing radioactive carbon (^{14}C). After two hours and six hours, scientists cut a section of the stem and exposed it to photographic film to produce autoradiographs shown below:

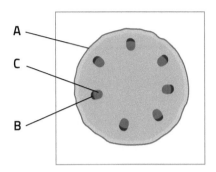

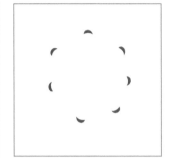

section of stem placed against photographic film in the dark

autoradiograph seen after two hours

autoradiograph seen after six hours

a) Identify structures A, B and C. (2)

b) Scientists concluded that sucrose containing ^{14}C was being translocated in the phloem. Explain how the evidence available supports that conclusion. (3)

Isla's answers

a) *A= endodermis* X
    B= Phloem ✓
    C= Xylem ✓
        2 correct = 1 mark

MARKER NOTE
Isla has confused endodermis with epidermis.

b) *It takes time for ^{14}C to be converted into glucose and then sucrose, which is why no image is seen after 2 hours.* ✓ *After 6 hours the autoradiograph shows dark areas* ✓ *in the same place where the phloem is found, showing that sucrose containing ^{14}C is being translocated* ✓*.*

Isla achieves 4/5 marks

Ceri's answers

a) *A= epidermis* ✓
    B= Phloem ✓
    C= Phloem X
        2 correct = 1 mark

MARKER NOTE
Ceri has confused the location of the xylem with phloem.

b) *There are some dark areas showing that radioactive sucrose must be moving through the phloem.* ✓

MARKER NOTE
Ceri has identified that dark areas are present but has made no attempt to use the evidence to explain why the results show it is the phloem, i.e. the dark areas are in the same place as the phloem. Ceri also needs to explain the different results for two and six hours.

Ceri achieves 2/5 marks

EXAM TIP
Always use all the available information to support your answer.

Section 4: Adaptations for nutrition

Topic summary

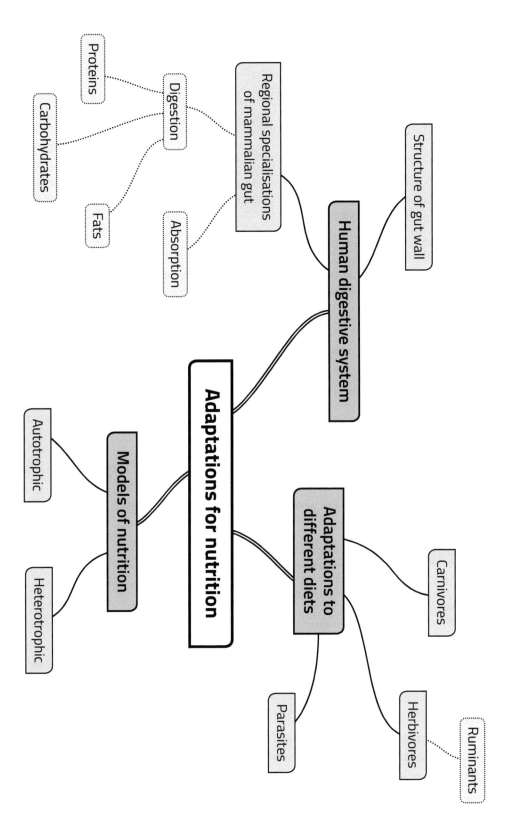

Practice questions

[AO1] [AO2]

Q1 The diagram below shows a longitudinal section through the alimentary canal:

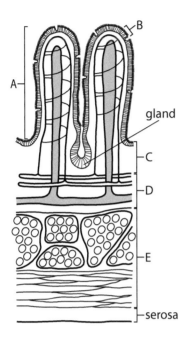

a) Use the diagram to complete the following table: (4)

Letter	Name	Function
B		Increases surface area of villus
C		Contains glands which produce mucus and enzymes
D	Submucosa	
E	Circular and longitudinal muscle layers	

b) Describe the function of the hepatic portal vein. (1)

....

....

c) Coeliac disease is a disease that affects the small intestine, causing the villi to become flattened. Suggest why symptoms often include diarrhoea and fatigue. (3)

....

....

....

....

....

[AO1] [AO2]

Q2 *Taenia solium* is an endoparasite that can affect humans. Its life cycle is shown in the diagram:

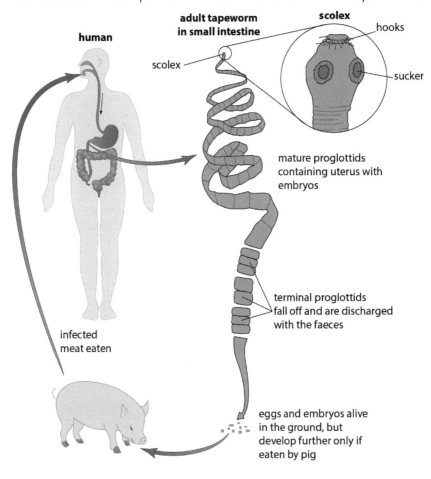

a) Define the term endoparasite. (1)

....

....

b) Explain **three** reproductive adaptations shown by *Taenia* **not shown in the diagram**. (3)

....

....

....

....

....

c) Praziquantel is a drug used to treat tapeworm infections, and is thought to work by the tapeworm losing the ability to resist digestion by the mammalian host. Using your knowledge of *Taenia*, suggest how the drug works to eliminate the parasite. (2)

....

....

....

[AO1]

Q3 Distinguish between the following modes of nutrition:

a) Autotrophic and heterotrophic nutrition (2)

....

....

....

....

b) Saprotrophic and holozoic nutrition (2)

....

....

....

....

c) Photoautotrophic and chemoautotrophic nutrition (2)

....

....

....

....

d) Excretion and egestion (2)

....

....

....

....

[AO2] [AO3]

Q4 Describe how proteins are digested and absorbed in the human alimentary canal, detailing how each part of the alimentary canal is adapted to the process. (9 QER)

[M] [AO2]

Q5 The following picture was taken using a light microscope from a section of the mammalian alimentary canal:

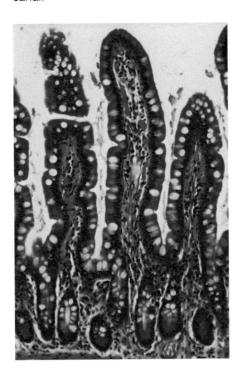

a) Identify the part of the alimentary canal shown. (1)

....

b) On the picture above, label the four tissue layers shown. (3)

c) The magnification of the image is 40×. Calculate the height of one villus. Show your working. (2)

Answer: ....

Question and mock answer analysis

Component 2 Practice questions

Q&A 1

[P] [M] [AO2] [AO3]

A student carried out an experiment to investigate the effect of temperature on the digestion and absorption of carbohydrates: 10 cm³ of a 1% w/v solution of starch was added to some Visking tubing (a semi permeable membrane which does not allow starch to pass through but does allow glucose to pass) which was sealed at both ends. The tubing was lowered into a water bath at 10°C and allowed to equilibrate for ten minutes. 1 cm³ of 1% amylase and maltase enzyme were carefully introduced through a tap in the top. 1 cm³ of water surrounding the tubing was removed every 20 seconds and tested for the presence of glucose by performing a Benedict's test. The time taken for the first appearance of glucose was recorded. The experiment was repeated for each temperature. The results are recorded below:

Apparatus

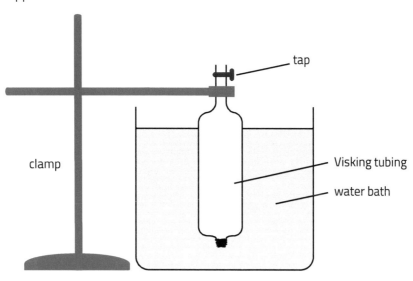

Results

Temperature /°C	Time taken for glucose to appear / s	Rate of digestion and absorption 1/time ×1000 s⁻¹
10	200	
20	120	
30	80	
40	40	
50	20	

a) Explain why the tube was left for ten minutes to equilibrate before adding the enzyme. (1)

b) Complete the table and draw a graph of the results to show the effect of temperature on the rate of digestion and absorption. (6)

c) The student concluded that the digestion and absorption rate doubled per 10°C rise. Evaluate this statement. (3)

Isla's answers

Temperature / °C	Time taken for glucose to appear / s	Rate of digestion and absorption 1/time ×1000 s⁻¹
10	200	5.0
20	120	8.3
30	80	12.5
40	40	25.0
50	20	50.0

✓

a) To ensure the contents of the tubing reached the same temperature as the water bath. ✓

b) ✓✓✓✓

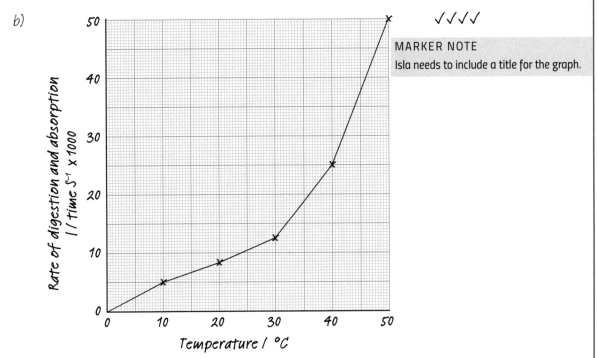

MARKER NOTE
Isla needs to include a title for the graph.

c) Rate only doubles per 10°C rise from 30°C to 50°C. Below this the increase is less than double. ✓ Some caution needs to be exercised as the conclusion is only based on one set of results. For greater confidence, the experiment should be repeated at least three times at each temperature so a mean can be calculated. ✓

MARKER NOTE
For a third mark Isla should refer to a major source of error being that samples of water are only removed for testing at 20-second intervals. It is quite likely that glucose could be detected between the intervals used; so to increase confidence, testing could be done more frequently, say at 10-second intervals.

Isla achieves 8/10 marks

Component 2 Practice questions

Ceri's answers

Temperature / °C	Time taken for glucose to appear / s	Rate of digestion and absorption 1/time ×1000 s⁻¹
10	200	5
20	120	8
30	80	12
40	40	25
50	20	50

X

MARKER NOTE

Calculation should be to at least 1 dp

a) To ensure the contents of the tubing were at the correct temperature. ✓

b)

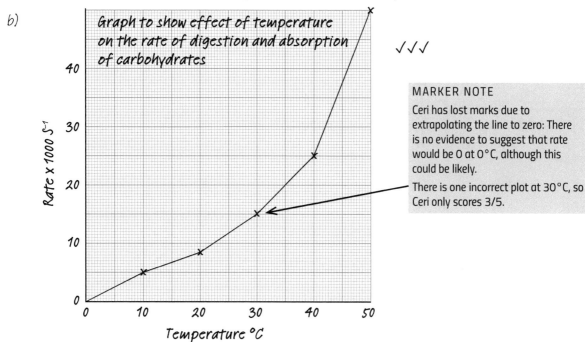

Graph to show effect of temperature on the rate of digestion and absorption of carbohydrates

✓✓✓

MARKER NOTE

Ceri has lost marks due to extrapolating the line to zero: There is no evidence to suggest that rate would be 0 at 0°C, although this could be likely.

There is one incorrect plot at 30°C, so Ceri only scores 3/5.

c) Rate only doubles per 10°C rise from 30°C to 50°C so the conclusion is correct in part. ✓ The experiment should be repeated at least three times at each temperature to increase accuracy of the experiment. X

MARKER NOTE

Increasing the number of repeats does not increase accuracy – it increases reliability and hence confidence in the conclusion if the repeats are similar.

Testing of the water samples should be made more frequently, say every five seconds, to improve confidence in the conclusion. ✓

MARKER NOTE

The mark was awarded here, but Ceri should have included the reason why, i.e. that glucose could be detected between the intervals used.

Ceri achieves 6/10 marks

EXAM TIP

Graphs usually attract five marks, so always remember your title, label axis with units, include an origin, ensure the axis is linear and take care with plotting points accurately, using a ruler to draw a straight line between them. You will lose one mark for each omission or error.

Practice papers

Component 1 Practice paper

75 Marks 90 minutes

1) The following diagrams represent the structure of common organic molecules:

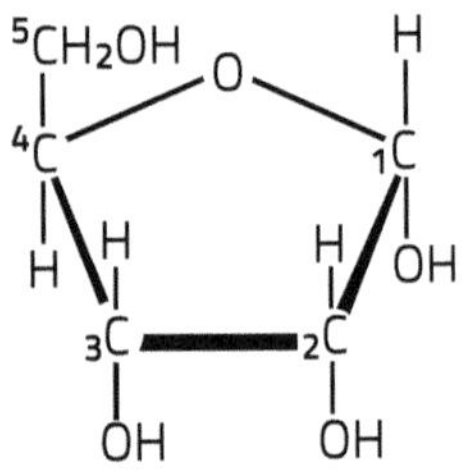

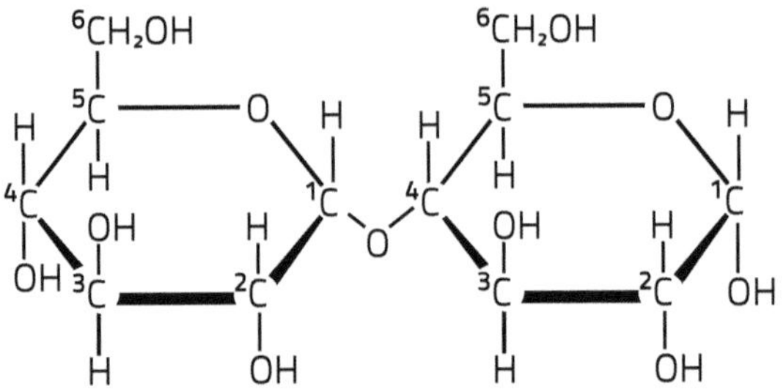

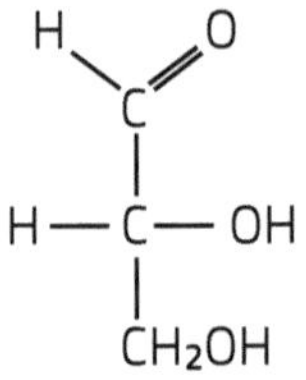

Using letters A–G, complete the table below. You may use each letter once, more than once or not at all.

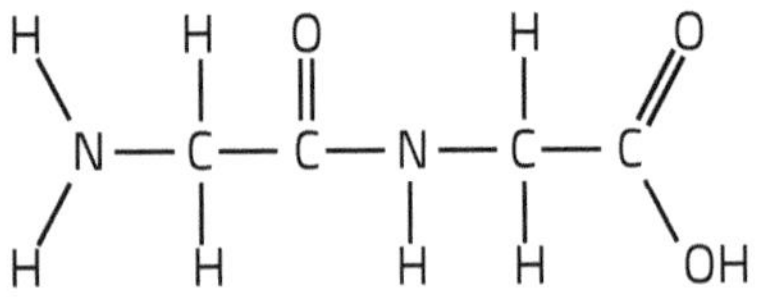

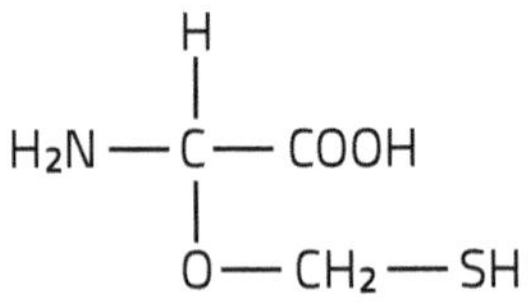

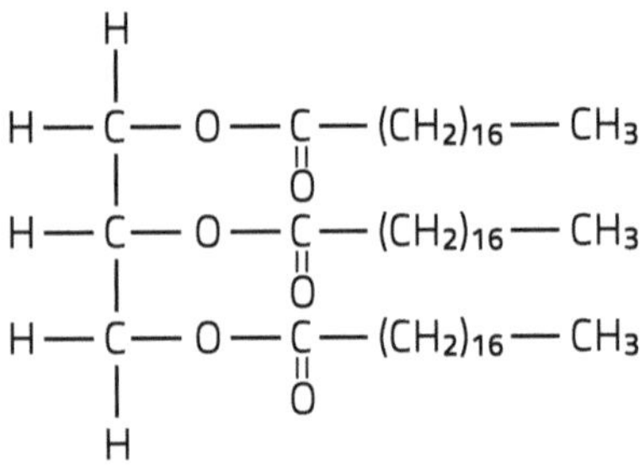

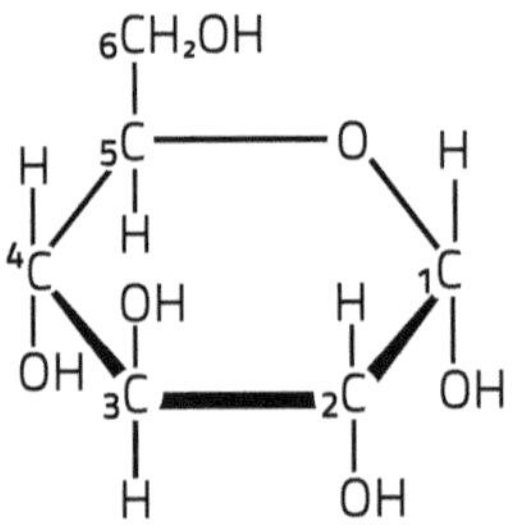

Statement	Letter(s)
Would be found in nucleic acids	
May contain C=C bonds	
Contains a glycosidic bond	
Is a triose sugar	

(4 marks total)

2) HIV is the virus which causes acquired immunodeficiency syndrome (AIDS). The virus attacks and destroys the CD4 lymphocytes which coordinate the immune response. HIV contains RNA as its nuclear material contained within a capsid, and reverse transcriptase enzyme, which converts the HIV RNA into DNA so the nuclear material can enter the CD4 cell nucleus and combine with the cell's nuclear material. Non-nucleoside reverse transcriptase inhibitors (NNRTIs) directly inhibit the HIV reverse transcriptase by binding in a reversible and non-competitive manner to the enzyme.

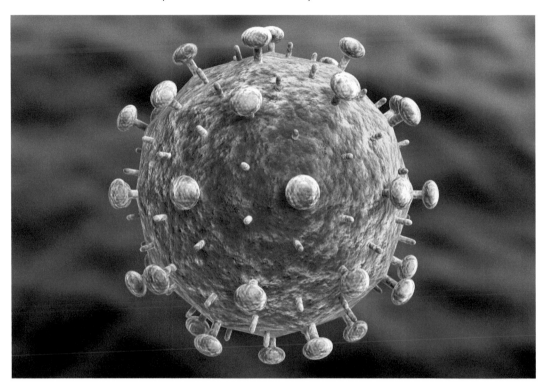

a) Draw a labelled diagram of a RNA nucleotide in the space below. (2)

b) Explain how a polynucleotide of RNA differs from DNA. (3)

....

....

....

....

....

....

c) The magnification of the above electron micrograph is 700,000×. Calculate the actual width of the HIV virus particle. Show your working. (2)

Answer: ....

d) Using your knowledge and the information above, answer the following questions:

i) Describe how HIV differs from bacteria such as *E. coli*. (2)

....

....

....

....

ii) Suggest how NNRTIs could prevent HIV from infecting CD4 lymphocytes. (2)

....

....

....

iii) Suggest why higher levels of virus present do not reduce the effectiveness of NNRTIs in preventing reverse transcription. (2)

....

....

....

....

(13 marks total)

3) Casein is a protein found in powdered milk which, when hydrolysed by the protease enzyme trypsin, turns from an opaque suspension to translucent. A student carried out an experiment to investigate the effect of pH on the rate of the enzyme-controlled reaction using a range of five different buffers from pH 6–10. The student timed how long it took for the milk to turn translucent at each pH, and recorded the results in the table below:

pH	Time taken for solution to become translucent/s				Rate of reaction / $s^{-1} \times 10^{-3}$
	Trial 1	Trial 2	Trial 3	Mean	Mean
6	433	436	420		
7	201	220	156		
8	130	119	150		
9	91	120	92		
10	358	355	378		

a)   Complete the table to show the mean time taken and mean rate of reaction. (2)

b)   Draw a graph to show the effect of pH on mean rate of reaction. (5)

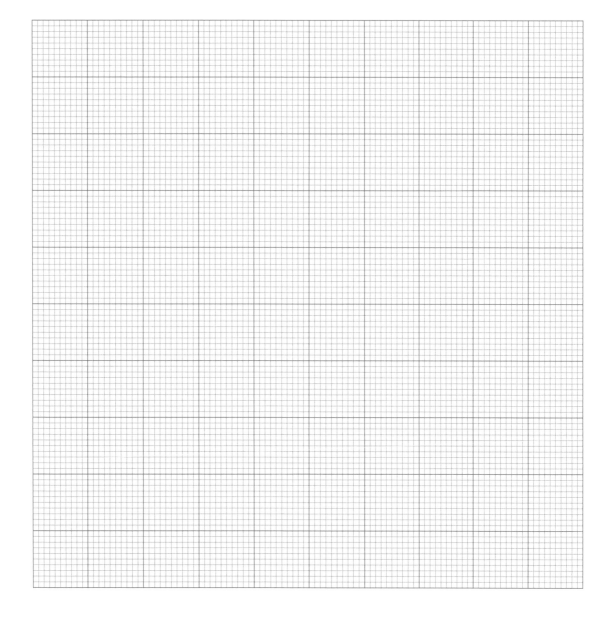

c)   Using the results and your knowledge draw a valid conclusion. (3)

....

....

....

....

....

d)   Discuss the accuracy of the experiment and suggest improvements which could be made. (3)

....

....

....

....

....

(13 marks total)

4) Plasma membranes are made up of phospholipids.

   a) In the space below, draw a labelled diagram to show the typical arrangement and properties of these components. (3)

   b) Describe how a phospholipid differs from triglyceride. (2)

   ....

   ....

   ....

   ....

c) In an experiment to determine the structure of the plasma membrane, scientists labelled the membrane proteins from two different cells using coloured dyes. A cell taken from a mouse had its membrane proteins labelled with a red dye, whilst a human cell had its membrane proteins labelled with a green dye. The two cells were then fused and the hybrid cell viewed immediately after fusion and again after one hour. The results are shown below:

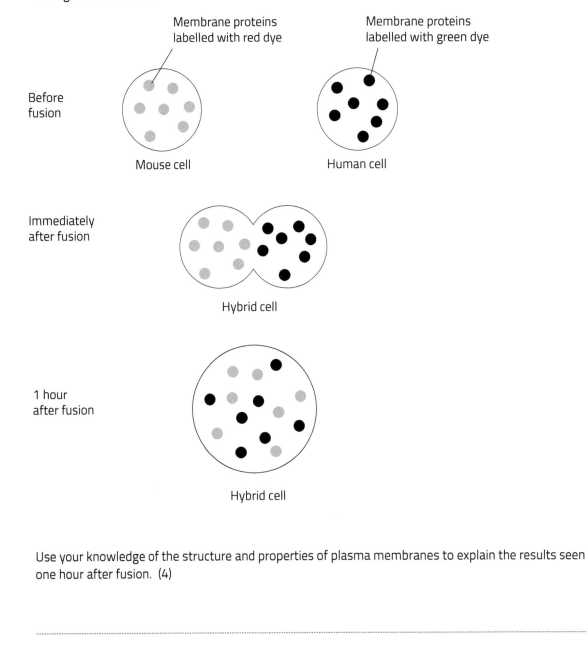

Use your knowledge of the structure and properties of plasma membranes to explain the results seen one hour after fusion. (4)

....

....

....

....

....

....

(9 marks total)

5) The following drawing was made by a student and shows four cells undergoing a form of cell division:

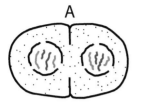
A

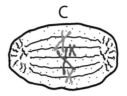
B

C

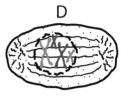
D

a) The student identified the type of cell division seen as being mitosis. What evidence is seen in the diagrams to support this conclusion? (3)

....

....

....

....

....

b) Identify stages A–D. (3)

A ....

B ....

C ....

D ....

c) Describe the events that take place between the stage shown in drawings C and B. (3)

....

....

....

....

(9 marks total)

6) A student carried out an experiment to estimate the molarity of onion epidermal cells. Onion epidermis was removed and mounted onto a microscope slide. Five drops of 0.5M sucrose solution was added and left for 15 minutes. The slide was then viewed under the light microscope and a drawing made of the cells seen; fifteen of the cells below were recorded as being plasmolysed.

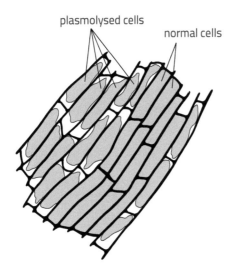

plasmolysed cells

normal cells

The student concluded that the molarity of the onion epidermal cells was less than 0.5M.

a) What evidence is there to support that conclusion? (3)

....

....

....

....

....

....

b) Outline how the experiment could be improved in order to estimate the actual solute potential of the onion cells. (4)

....

....

....

....

....

....

(7 marks total)

6) Sickle-cell anaemia is a disease caused by the alteration of a single nucleotide in the gene for the beta chain of the haemoglobin protein. Beta haemoglobin is a single chain of 147 amino acids, and in patients suffering from the disease, a single-base mutation results in the beta chain being deformed. This substitution is depicted in Table 1.

Table 1: Single-base mutation associated with sickle-cell anaemia

Sequence for normal haemoglobin												
ATG	GTG	CAC	CTG	ACT	CCT	GAG	GAG	AAG	TCT	GCC	GTT	ACT

Sequence for sickle-cell haemoglobin												
ATG	GTG	CAC	CTG	ACT	CCT	GTG	GAG	AAG	TCT	GCC	GTT	ACT
START	Val											

Adapted from Clancy, S. (2008) Genetic mutation. *Nature Education* 1(1):187

a) What is meant by a codon? (2)

....

....

b) Complete the table above to show the amino acid sequence for sickle-cell haemoglobin. The first two have been done for you. (3)

c) Which amino acid is changed as a result of the single-base mutation? (1)

....

d) Distinguish between introns and exons, and explain how they are removed during transcription. (5)

....

....

....

....

....

....

....

(11 marks total)

7) The endosymbiotic hypothesis for the origin of mitochondria and chloroplasts suggests that the organelles are descended from specialised bacteria which were engulfed by a cell and became incorporated into the cytoplasm. Using your knowledge of the structure of prokaryotes, mitochondria and chloroplasts, evaluate this theory. (9 QER)

Component 2 Practice Paper

Component 2 Practice paper

75 Marks 90 minutes

1)   The following diagrams show skulls from two different heterotrophic animals:

Animal 1

Animal 2

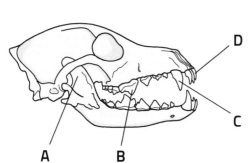
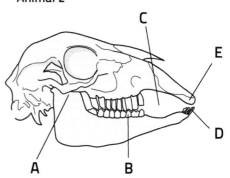

a)   Explain what is meant by heterotrophic nutrition. (1)

....

....

b)   Using your knowledge and the diagrams, answer the following questions:

   i)   Identify the type of nutrition each animal is adapted to. (2)

       Animal 1 = ...

       Animal 2 = ...

   ii)   Identify feature B in animal 1. (1)

....

....

   iii)   Explain how feature A is adapted in both animals to enable them to eat their food. (2)

....

....

....

c)   The length of the colon is much longer in animal 2 than animal 1. Explain why. (3)

....

....

....

(9 marks total)

2) An ECG was taken at rest and following moderate exercise.

At rest   Moderate exercise

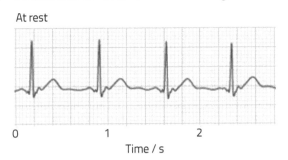

a) What is meant by an ECG? (1)

....

....

b) Calculate the average heart rate at rest and following moderate exercise. Show your working. (3)

At rest = ...

After moderate exercise = ...

c) Apart from the heart rate, describe one other change that occurs in the ECG following moderate exercise, and explain its significance. (2)

....

....

....

(6 marks total)

3) The diagram shows an alveolus from the human lungs:

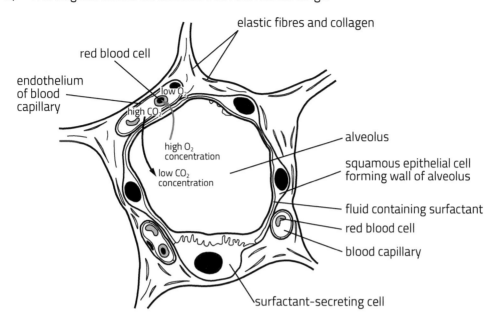

a) Explain how alveoli are adapted for gas exchange. (4))

....

....

....

....

....

....

....

....

b) Explain why premature babies born before 23 weeks of pregnancy require use of artificial surfactant. (3)

....

....

....

....

....

(7 marks total)

4) Anaemia is a condition where the number of red blood cells or the level of haemoglobin is lower than normal, and can result in tiredness and inability to do strenuous exercise. It can be caused by a number of conditions including a lack of iron.

a) Using your knowledge of the structure of haemoglobin, explain why a lack of iron could cause the symptoms of anaemia. (3)

....

....

....

....

b) When suffering from chronic anaemia, the oxyhaemoglobin dissociation curve is shifted to the right, as shown in the dissociation curve below:

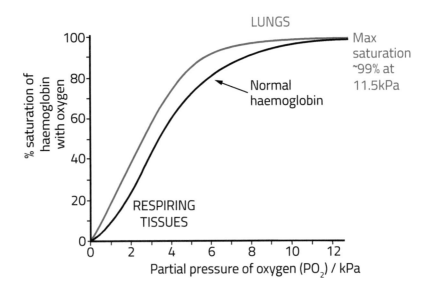

Explain the significance of this shift in oxygen dissociation to the patient. (3)

....

....

....

....

c) On the graph, draw the dissociation curve for the lugworm, *Arenicola marina* which lives in muddy burrows where oxygen availability is low. Explain the position of your line. (4)

....

....

....

....

....

(10 marks total)

5) Species diversity is used as an indicator of freshwater pollution. Scientists conducted kick sampling at two different sites in a river to assess the number and type of freshwater invertebrates present in order to calculate the Simpson's diversity Index (S). The following results were obtained:

Species	Number present near factory outlet (Site A)	Site A $n(n-1)$	Number present 2 miles downstream of factory (Site B)	Site B $n(n-1)$
Stonefly nymph	0		15	
Mayfly larva	0		12	
Freshwater shrimp	0		11	
Caddis fly larva	3		4	
Bloodworm	7		2	
Water louse	9		0	
Red-tailed maggot	11		0	
	N = N−1 =	$\Sigma n(n-1) =$	N = N−1 =	$\Sigma n(n-1) =$

a) Define the term biodiversity. (1)

....

....

b) Describe how sampling of both sites should be carried out including any precautions used to ensure results are reliable. (3)

....

....

....

....

c) Calculate N, N−1 and S n(n−1) for each site and enter your answers in the table above. Using the Simpson's diversity index formula below, calculate S for each site. (4)

$$S = 1 - \frac{\sum n(n-1)}{N(N-1)}$$

S (site A) = ...

S (site B) = ...

d) Scientists concluded that the Simpson's diversity index was not a good way of measuring impact of freshwater pollution, and instead the absence of certain species, called indicator species, was more reliable. Using the results, assess the validity of this conclusion. (4)

....

....

....

....

....

....

(12 marks total)

6) The dissolved oxygen level in water (mg/L) varies greatly with temperature and where water is fast-moving, it generally has more oxygen than still water, because the movement mixes oxygen from the air into the water. In the UK, freshwater temperature varies between 2–7°C during winter and 8–16°C during summer and is influenced by other factors such as altitude, depth and turbidity (cloudiness).

Table 1 Dissolved oxygen saturation at different water temperatures

Temperature / °C	Dissolved oxygen saturation (mg/L) in still water
0	14.6
5	12.8
10	11.3
15	10.1
20	9.1
25	8.3
30	7.6

Table 2 Minimum oxygen requirements by fish species

Fish species	Minimum dissolved oxygen requirements (mg/L)
Trout	13–14
Salmon	11–12
Bass	9–10
Carp	7–9
Walleye	5–7
Pike	3–5

a) Using your knowledge and the information provided, answer the following questions.

i) Suggest why trout are usually found in fast-moving rivers in the UK. (3)

ii) Suggest why pike can be found in lakes and rivers in the UK. (2)

iii) Suggest why bony fish are able to tolerate lower dissolved oxygen saturation levels than cartilaginous fish. (4)

(9 marks total)

7) A student investigating the effect of air movement on transpiration in privet leaves set up the apparatus as shown below with all joints sealed with Vaseline. A desk fan with three different speeds was used to vary the air speed, and an anemometer to record the actual wind speed in ms^{-1}. The distance the bubble moved in one minute was recorded. The capillary tube had a diameter of 1 mm. The results are shown below.

Apparatus

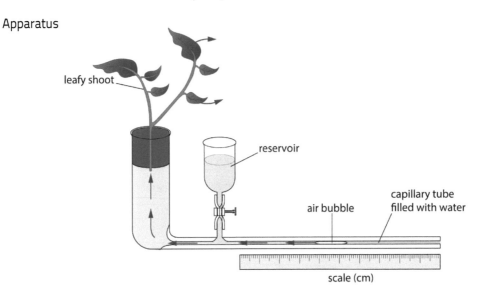

Air speed / ms^{-1}	Distance moved by the bubble / mm	Distance moved by the bubble / mm	Distance moved by the bubble / mm	Mean distance moved by the bubble / mm	Mean volume of water taken in / mm^3 min^{-1}
0	2	2	3	2.3	
5	12	11	13	12.0	
10	39	22	30	30.3	
15	45	42	48	45.0	

a) Calculate the mean volume of water taken in / mm^3 min^{-1} and write your answer in the table above. Show your working below using the formula volume = $\pi r^2 d$, where π =3.14, r = radius of capillary tube and d = distance bubble moved. (3)

Answer: ....

b) The student concluded that doubling the air speed doubled the rate of transpiration. Evaluate this statement. (3)

....

....

....

....

....

c)   Outline how the procedure could be improved. (4)

....

....

....

....

....

....

d)   Explain how the results would have been different if a xerophyte such as marram grass was used
instead of privet. (3)

....

....

....

....

....

(13 marks total)

8) In biology, structure is related to function. With reference to blood vessels, evaluate this statement. (9 QER).

Answers

Practice questions:
Component 1: Basic Biochemistry and Cell Organisation

Section 1: Chemical elements and biological compounds

Q1

a) $-H_2O$ shown

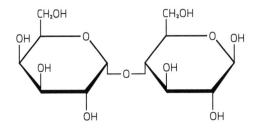

b) Condensation

c) Lactose, (1-4) glycosidic

d) A reducing sugar is <u>able to donate an electron to reduce another compound,</u> e.g. copper II sulphate to copper I oxide.

Q2 a) Hydrogen bond

b) +ve charge shown on H and negative on O, i.e.

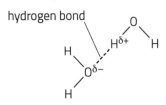

c) Water molecules attract each other (1)
and hydrogen bonds form between molecules (1)
Ref cohesion tension / attraction between water molecules and wall of xylem (1)

Q3 a) A = Cellulose, B = Chitin (1)

b) Answer MUST be comparative
<u>Both</u> are made of <u>many beta</u> glucose molecules joined together (1)
with alternate molecules rotated through 180° (1)
<u>Both</u> have long parallel chains cross-linked by hydrogen bonds into microfibrils (1)
B/Chitin has acetylamine groups / contains nitrogen which A/cellulose doesn't (1)
A/Cellulose has gaps between fibres in plant cell walls which are freely permeable to water, B/Chitin is more waterproof (1)
<u>Both</u> molecules convey strength (1)
MAX 5

Q4 a) Lipids

b) Circle drawn around O–H on carboxyl group (COOH)

c) Answer MUST be comparative
A is a saturated fatty acid whereas B is unsaturated (1)
A has no C=C (double-carbon bond) whereas B has one C=C bond / A has more hydrogen atoms than molecule B (1)
A occurs as fats which are semi-solid at room temperature whereas B occurs as oils which are liquid at room temperature (1)
high number of C–H bonds (more than carbohydrates) mean they both liberate twice as much energy as carbohydrates and function as energy reserves in plants and animals / saturated fats are useful stores of fats in mammals whereas oils are useful stores in plants (1)
MAX 3

Q5

a) Bond A = peptide bond (1)

```
    H     R₁    O              H     R₂    O
     \    |    //              \    |    //
      N — C — C                 N — C — C
     /    |    \               /    |    \
    H     H    OH      H       H     H    OH
                   ↑
                 + H₂0
```

Diagram MUST show chemical addition of water (1)
And both amino acids correctly drawn (1)

b) Peak enzyme activity for enzyme C is 55°C AND enzyme C has the highest sulphur composition / is 18.6% (1)
This would enable more disulphide bonds to form (1)
Disulphide bonds are the strongest bonds in tertiary structure (1)
so enzyme {less likely to be <u>inactivated</u> / <u>denatured</u> by heat} (1)

Q6

a) Solution A contains reducing sugar but no protein (1)
Solution B contains no reducing sugar but contains protein (1)
Solution C contains a low concentration of reducing sugar and protein (1)

b) The non-reducing sugar present has been hydrolysed (1)
to form a reducing sugar (1)

c) Test equal volumes of solution B and C (1)
use equal {volumes / concentrations} of biuret reagent (1)
compare the intensity of colour / use a colorimeter (1)
deeper colour shows higher protein concentration) (1)

Q7

a) Reference to α-helix / β-pleated sheet (1)
held together by hydrogen bonds (1)

b) Temperature has no effect on the length of the primary structure (1)
Increasing temperature affects the secondary structure resulting in an increase in length at 60 °C (1)
Correct manipulation of figures, e.g. 62% longer (1)

c) Reference to increase in kinetic energy at (60 °C) (1)
increased vibrations within molecule, breaking hydrogen bonds holding α-helix / β-pleated sheet together (1)
causing the molecule to unwind (1)

Q8

a) Sequence / order of amino acids (1)

b) Hydrogen bonds (1)

c)

Collagen	Haemoglobin
Only made of polypeptide chains	Made of polypeptide chains and a prosthetic haem group
Fibrous protein	*Globular protein*
Structural protein	*Transport protein / carries oxygen*
Consists of one type of polypeptide chain	*Two different polypeptide chains (2 alpha and 2 beta)*

Section 2:
Cell structure and organisation

Q1

a) A = Golgi body, B = rough endoplasmic reticulum (NOT RER) (1)

b) B has ribosomes whereas A does not (1)

c) Both are involved in transport (1)
B is involved in protein synthesis / translation whereas A is involved in assembling glycoproteins (1)
B is involved in packaging whereas A is involved in storage (1)
MAX 2

Q2 a) Nucleus and basement membrane (1)

b) Ciliated columnar epithelium (1) found in lining {bronchi allow trachea/uterus} (1)

c) Cilium allow cilia (1) {remove mucus from bronchi allow trachea / waft secondary oocyte (allow ovum) along oviduct} (1)

Q3 a)

Typical animal cell	Typical prokaryotic cell
No cell wall	**Cell wall made of peptidoglycan**
No chloroplasts	No chloroplasts
80 S ribosomes	70 S ribosomes
Ribosomes attached to membranes	Ribosomes loose in cytoplasm
Mitochondria	**Mesosome**
Nucleolus	No nucleolus
No plasmid	May contain plasmids

b) All four correct labels = 3 marks, three correct labels = 2, two correct = 1

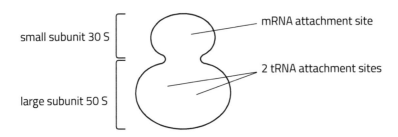

small subunit 30 S

large subunit 50 S

mRNA attachment site

2 tRNA attachment sites

Q4 a) Cold to reduce enzyme activity (1)
isotonic to reduce damage caused by osmosis / reduce osmotic lysis (1)
buffer to resist any changes in pH / maintain pH (1)
Any two

b) Nuclei are the largest organelle, ribosomes the smallest (1)
lysosomes and vesicles are approximately the same size (1)

c) Add the pellet to an isotonic solution of glucose (1)
look for presence of carbon dioxide indicating aerobic respiration (1)

Q5 a)

Feature	Name	Function
A	DNA	Site of transcription / DNA replication
B	Mesosome	Site of respiration
C	Ribosome (70S)	Site of translation
D	Plasmid	Confers antibiotic resistance

1 mark for each row

b) In eukaryotic cells C would be membrane bound (1), and would be larger 80S (1)

Q6 a) Chloroplast (1)

b)

Feature	Name	Function
A	(70S) ribosome	Site of translation
B	thylakoid	Site of the light-dependent reaction allow photosynthesis
C	DNA	Site of transcription / DNA replication
D	stroma	Site of the light-independent reaction / Calvin cycle

c) Inner membrane is folded in mitochondrion to form cristae (1)
mitochondria do not contain {starch grains/chlorophyll} (1)

Q7 a) An organ is an aggregation of several tissues (1)
to carry out a particular function (for the whole organism) (1)

b) Magnification is how many times bigger the image is compared to the object (1)
whilst the resolving power is the minimum distance by which two points must be separated in order
for them to be seen as two distinct points rather than a single focused image (1)

Q8 a) 2 marks for any 3 correct labels shown, 1 mark for any 2 correct labels
1 mark for diagram with at least three flattened sacs (cisternae) and one vesicle

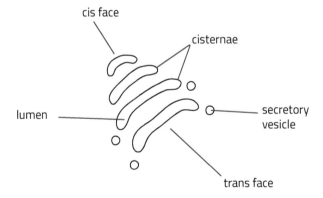

b) Rough endoplasmic reticulum has ribosomes whereas the Golgi body does not (1)
<u>both</u> are involved in transport (1)
Rough endoplasmic reticulum is involved in protein synthesis / translation whereas Golgi is involved in
assembling glycoproteins (1)
Rough endoplasmic reticulum is involved in packaging whereas Golgi is involved in storage (1)

Section 3:
Cell membranes and transport

Q1 a) A = phosphate head
B = fatty acid tail (1)
C = cholesterol (1)
all 3 for 2 marks
2 correct for 1 mark

b) Polar molecules pass across the membrane by facilitated diffusion (1)
using channel proteins (1)
Ref to active transport / carrier protein negates mark above as this is not passive movement
Diffusion is proportional to $\dfrac{\text{surface area} \times \text{concentration gradient}}{\text{diffusion distance}}$ (1)
{membrane is thin /two phospholipids thick} which reduces diffusion distance (1)

Q2 a) Graph with title and axis labelled (1)
all plots correct (2) – 1 mark for each incorrect plot
Axis correct with origin (1)
Both lines drawn point to point (1)
–1 if scale poor, i.e. not using at least ¾ of paper

Rate of oxygen uptake at 20°C and 40°C

e.g.

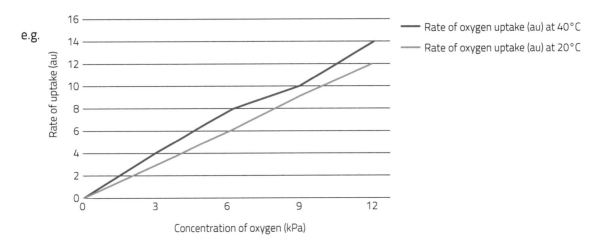

b) Rate of oxygen uptake is proportional to oxygen concentration (1)
Data to support results (1)
this suggests method of uptake is diffusion (1)
Doubling temperature increases rate of uptake (but does not double it) (1)
increasing temperature will increase vibration of molecules and rate of diffusion (1)

c) Add a respiratory inhibitor, e.g. cyanide (1)
If active transport was occurring, rate of uptake would drop rapidly after adding it (1)

Q3 a) All calculations correct 2 marks. –1 for each incorrect.

Molarity of sucrose solution (M)	Number of cells plasmolysed	Total number of cells observed	Percentage plasmolysis %
0.0	0	70	0.0
0.1	19	75	25.3
0.2	28	65	43.1
0.3	37	66	56.1
0.4	49	72	68.1
0.5	65	71	91.5

b) Graph with title and axis labelled (1)
all plots correct (2) – 1 mark for each incorrect plot
Axis correct with origin (1)
line drawn point to point (1)
–1 if scale poor, i.e. not using at least ¾ of paper
e.g.

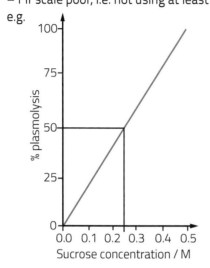

Incipient plasmolysis from graph = point at which 50% of cells are plasmolysed (between 0.2 and 0.3M) (1)

c) Correct value with units, e.g. –680 kPa to nearest value (1)

d) Water potential = solute potential (1) and
pressure potential = 0 (1)

Q4 a) (i) By <u>facilitated</u> diffusion (1) as it moves {with the concentration gradient / from high concentration to low concentration} (1) from 42.0 mmol L^{-1} to 5.1 mmol L^{-1} (1)

(ii) By active transport (1) as it moves {against the concentration gradient / from low concentration to high concentration} from 4.9 mmol L^{-1} to 130.0 mmol L^{-1} (1)
must include units for third marking point

b) Data supports this as glucose movement is {against the concentration gradient / from low concentration to high concentration} as it is moving from 12.7 mmol L^{-1} to 42.0 mmol L^{-1} (1)
This is powered by co-transport of sodium ions {down their concentration gradient/from high to low concentration/from 83.9 mmol L^{-1} to 4.9 mmol L^{-1}} (1)
However, glucose could be entering by active transport (1)

Q5 a) <u>Facilitated</u> diffusion (1)

b) (Ions are not transported into mucus)
so solute potential of mucus is not lowered (1)
water is not drawn out of cells into mucus (1)
<u>by osmosis</u> thinning it (1)
ignore refs to water potential

c) Inhale sodium chloride solution/saline (1)
to lower solute potential of mucus (1)
Allow:
inhale bronchodilator, e.g. Ventolin (1)
to widen bronchioles (1)

Q6 d) Bacterium engulfed into vesicle (1)
Lysosomes labelled and shown emptying
enzymes into vesicle (1)
Bacterium digested (1)

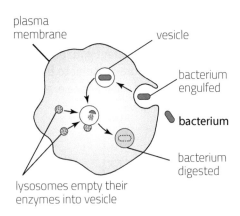

plasma membrane
vesicle
bacterium engulfed
bacterium
bacterium digested
lysosomes empty their enzymes into vesicle

b) Solid materials enter by phagocytosis (1)
Liquid materials enter by pinocytosis (1)
Both involve movement of material by cell
membrane folding inwards/surrounding material (1)
to form vesicles (1)
Vesicles are smaller in pinocytosis (1) allow converse
MAX 4

Section 4:
Enzymes and biological reactions

Q1 a) The minimum energy that must be put into a chemical system for a reaction to occur (1)

b) Enzymes lower the activation energy (1)
by providing energy to break bonds in existing ones so new ones can form in new molecules (or WTTE) (1)

c) Denatured is a permanent change whereas inactivated is reversible (1)
Inactivation occurs either side of the optimum pH, whereas denatured occurs more at the extremes (1)
Inactivation involves charges on active site repelling substrate whereas denaturation involves ionic bonds in active site breaking causing a permanent change to the shape of the active site (1)
Answer must be comparative

d) Correct shape (1)
Correct position with optimum at pH 7, fully denatured below pH 5 and above pH 9 (1) i.e.

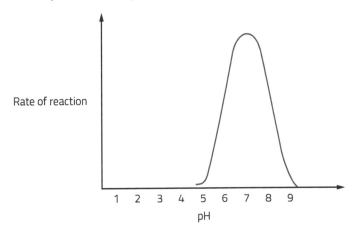

Rate of reaction

1 2 3 4 5 6 7 8 9
pH

Q2 a) Methanol and ethanol have similar chemical structures (1)
Ref to ethanol being competitive inhibitor of alcohol dehydrogenase (1)
so different product produced / no formaldehyde produced (1)
ref to no substrate for aldehyde dehydrogenase so no formate produced (1)
enzymes in cellular respiration are not inhibited (1)

b) Cyanide is a non-competitive inhibitor so binds irreversibly to an allosteric site (1)
whereas methanol binds temporarily to the active site (1)
the effect of methanol can be overcome by increasing the concentration of a competitive inhibitor/ ethanol whereas cyanide cannot (1)
No marks for reference to direct/indirect inhibition as it is in the question stem

You are not awarded a tick per point, but rather an assessment of your answer is made awarding three main bands. Awarding a mark within the band will depend on how fully you meet the statement.

7–9 marks
Indicative content of this level is...
Detailed description of a biosensor and how it works
Detailed explanation of three or more advantages of using immobilised enzymes

The candidate constructs an articulate, integrated account, correctly linking relevant points, such as those in the indicative content, which shows sequential reasoning. The answer fully addresses the question with no irrelevant inclusions or significant omissions. The candidate uses scientific conventions and vocabulary appropriately and accurately.

4–6 marks
Indicative content of this level is...
Description of a biosensor and how it works
Explanation of two advantages of using immobilised enzymes

The candidate constructs an account correctly linking some relevant points, such as those in the indicative content, showing some reasoning. The answer addresses the question with some omissions. The candidate usually uses scientific conventions and vocabulary appropriately and accurately.

1–3 marks
Indicative content of this level is...
Basic description of a biosensor and how it works
Basic explanation of one advantage of using immobilised enzymes

The candidate makes some relevant points, such as those in the indicative content, showing limited reasoning. The answer addresses the question with significant omissions. The candidate has limited use of scientific conventions and vocabulary.

0 marks
The candidate does not make any attempt or give a relevant answer worthy of credit.

A good answer would therefore include:
- Definition, i.e. biosensors contain immobilised enzymes that can be used to detect small concentrations of specific molecules in a mixture, e.g. glucose in a sample of blood.
- A biosensor consists of a specific immobilised enzyme, a selectively permeable membrane, and a transducer connected to a display.
- The selectively permeable membrane allows the metabolite to diffuse through to the immobilised enzyme, whilst preventing the passage of other molecules.
- The metabolite binds to the active site of the enzyme, and is converted into electrons, which in turn combine with the transducer turning the chemical energy into an electrical signal. The higher the concentration of metabolite present, the greater the electrical signal.
- Immobilised enzymes are enzymes that are fixed to an inert matrix, e.g. entrapment – held inside a gel, e.g. silica gel, or by micro-encapsulation – trapped inside a micro-capsule, e.g. alginate beads.

Advantages
- Small concentrations of <u>specific</u> molecules, i.e. glucose can be detected.
- Enzymes are contained within their own 'micro-environment', the enzymes are less susceptible to changes in pH, temperature.
- This technique is used to accurately measure the blood glucose of diabetic patients (whose blood glucose should normally be kept between 3.89 and 5.83 mmol dm^{-3}).
- Product is not contaminated by the enzyme.
- Can use different enzymes, e.g. ref to GOx and GDH, with different temperature and pH optima.

Q4
a) Product not contaminated with enzyme (1)
enzyme can tolerate wider range of temperatures/pHs (1)

b) Hydrolysis (1)

c) Increasing the flow rate increases the concentration of substrate up to 5 cm³ min⁻¹ (1)
above 5 cm³ min⁻¹ {no further increase in rate of glucose production/ glucose production rate remains constant} (1)
fastest rate of production of glucose is between 0 and 6 cm³ min⁻¹ (1)
this is because all active sites are full (1)

d) Reduce the size of the beads to increase surface area of enzymes available (1)
increase the temperature to increase the kinetic energy of enzyme and substrate molecules (1)
to increase chance of successful collisions (1)

Q5
a) Any value between 30 and 35°C (1)

b) All three enzymes have no activity above 50°C so would not remove protein stains (1)
because they are denatured (1)
hydrogen bonds in tertiary structure have broken (1)
so active site is deformed and substrate molecules no longer fit (1)
Any 3

c) Each enzyme has a specific shaped active site (1)
which will only fit a specific substrate / ref to egg, milk, blood being different shaped substrates (1)

d) Enzyme EC101 which removes blood stains {has been denatured / activity is zero} (1)
other enzymes (which are working) don't remove blood stains (1)

Section 5:
Nucleic acids and their functions

Q1
a) % A is approximately equal to %T AND %G is approximately equal to %C (1)
due to complementary base pair on antisense strand (1)

b) % A won't equal %T AND %G won't equal %C (1)
as there is no complementary base pairing due to being single-stranded (1)

c) 3.4 / 0.34 = 10 (1)
×3 = 30 (1)

Q2
a) A = Phosphate
B = Deoxyribose (not pentose)
C = Nitrogenous base
all 3 correct = 2 marks
one incorrect =1 mark

b) RNA single-stranded, DNA double-stranded (1)
RNA would have base uracil instead of thymine (1)
need comparison

c) clover leaf shape (1)
correct position and label for anticodon (1)
showing some hydrogen bonds (1)   e.g.

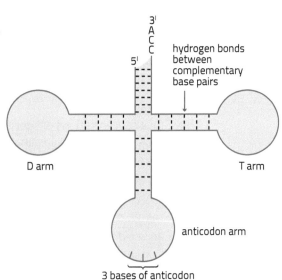

D arm

3'
A
C
C
5'

hydrogen bonds between complementary base pairs

T arm

anticodon arm

3 bases of anticodon

Q3 a) Two bands drawn, one in light position, other intermediate (1)

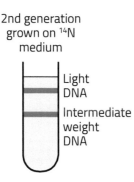

2nd generation
grown on ¹⁴N
medium

Light DNA

Intermediate weight DNA

b) Intermediate weight band seen (1)
indicating the DNA molecule contained one strand from the heavy parent DNA and one newly synthesised light DNA strand. (1)

c) The DNA extracted formed an intermediate weight band halfway up the tube, and a lighter band towards the top of the tube (1)
Because half of the DNA was intermediate weight and half light, this rules out dispersive replication (1)

d) The proportion of light DNA produced would increase/ light band would become thicker (1)

Q4 a) One mark per correct row (3)

Organism	Base composition (%)			
	A	T	G	C
Yeast	31.3	**31.3**	**18.7**	**18.7**
Rat	**28.4**	28.4	**21.6**	**21.6**
Human	30.1	**30.1**	19.9	**19.9**

b) Complementary base pairing rule wound not apply / A≠T and G≠C (1)

c) Three H–H bonds between G–C, only two between A–T (1)
more hydrogen bonds in double helix would require more energy to break and denature molecule (1)

Q5 a) Transcription – nucleus
Translation – ribosome

b) DNA 5' AAA AGA TGA GCA TCA CCT CTT 3'
mRNA **UUU UCU ACU CGU AGU GGA GAA**
−1 for each incorrect triplet

c) Phenylalanine, serine, threonine, cysteine, serine, glycine, glutamic acid
−1 for each incorrect amino acid

Q6 You are not awarded a tick per point, but rather an assessment of your answer is made awarding three main bands. Awarding a mark within the band will depend on how fully you meet the statement.

7–9 marks
Indicative content of this level is...
Detailed description of steps in transcription
Detailed description of steps in translation
Detailed description of steps in posttranslational modification

The candidate constructs an articulate, integrated account, correctly linking relevant points, such as those in the indicative content, which shows sequential reasoning. The answer fully addresses the question with no irrelevant inclusions or significant omissions. The candidate uses scientific conventions and vocabulary appropriately and accurately.

4–6 marks
Indicative content of this level is...
Description of steps in transcription
Description of steps in translation
Description of steps in posttranslational modification

The candidate constructs an account correctly linking some relevant points, such as those in the indicative content, showing some reasoning. The answer addresses the question with some omissions. The candidate usually uses scientific conventions and vocabulary appropriately and accurately.

1–3 marks
Indicative content of this level is...
Basic description of steps in transcription
Basic description of steps in translation
Basic description of steps in posttranslational modification

The candidate makes some relevant points, such as those in the indicative content, showing limited reasoning. The answer addresses the question with significant omissions. The candidate has limited use of scientific conventions and vocabulary.

0 marks
The candidate does not make any attempt or give a relevant answer worthy of credit.

A good answer would therefore include:
- DNA acts as a template for the production of mRNA
- Description of DNA helicase action
- Free RNA nucleotides pair and action of RNA polymerase
- Ref STOP codon
- In eukaryotes, pre-mRNA is spliced to remove the intron non-coding regions before passing to the ribosomes
- mRNA strand leaves the nucleus via the nuclear pores and moves to the ribosomes
- Translation involves transfer RNA
- Ref exposed bases called the anticodon, which are complementary to the mRNA codon
- Attachment of the relevant amino acid to the attachment site is called amino acid activation and requires ATP
- Translation involves converting the codons on the mRNA into a sequence of amino acids known as a polypeptide
- Description of process involved
- Translation produces a polypeptide, but further modification is needed in order to produce a protein with a secondary, tertiary or quaternary structure
- Accept the suggestion that modification occurs within the Golgi body
- To form haemoglobin, two alpha chains and two beta chains (coded by two different genes) need to be assembled together with iron as a prosthetic group.

 a) All three parts labelled correctly for 2 marks
two parts = 1 mark

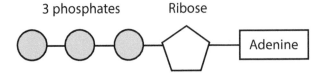

b) The high energy bond between 2nd and 3rd phosphate group is broken by hydrolysis to produce ADP and (1)

releases 30.6 kJ of energy to the cell (1)

c) active transport (1)

muscle contraction (1)

nerve impulse transmission (1)

used in anabolic reactions, e.g. DNA/protein synthesis (1)

MAX 2

d) Energy is released quickly from a one-step process involving just one enzyme (1)

energy is released in small amounts (1)

Is universal energy currency / common source of energy for all reactions in all living things (1)

Section 6:
Cell cycle and cell division

Q1 a) An allele is one form of the same gene (1)

which is a section of DNA coding for a particular polypeptide (1)

b)

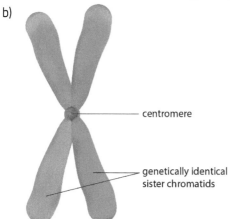

centromere

genetically identical
sister chromatids

1 mark for X shape

1 mark for both centromere and chromatids correctly labelled

c) i) An organism with two complete sets of chromosomes

ii) *Aedes agypti* = C (1) and D (1)
Chagasia bathana = B (1)

iii) 4 different chromosomes drawn i.e.

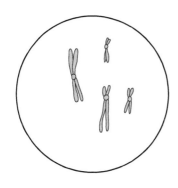

Q2
a) A (late) Anaphase
   B (early) Anaphase
   C Telophase
   D Prophase
   E Metaphase
   All 5 correct for 3 marks, -1 for each incorrect

b) Idea that A is later in anaphase / chromosomes have been pulled more towards each pole (1)
   due to spindle fibres shortening (1)

c) i)

Root tip sample	Number of cells seen in				% cells undergoing mitosis
	Prophase	Metaphase	Anaphase	Telophase	
1	12	5	3	7	
2	21	6	2	8	
3	18	3	2	9	
Mean	**17.0**	**4.7**	**2.3**	**8.0**	

ii) Anaphase, metaphase, telophase, prophase
   all for correct = 2 marks
   1 incorrect = 1 mark

iii)

Root tip sample	Number of cells seen in				% cells undergoing mitosis
	Prophase	Metaphase	Anaphase	Telophase	
1	12	5	3	7	**22.5**
2	21	6	2	8	**30.8**
3	18	3	2	9	**26.7**
Mean	17	4.7	2.3	8.0	**21.9**

Whole column correct = 2, −1 for each incorrect

iv) The mitotic index (percentage of cells undergoing mitosis) is highest in cells that are more actively dividing (1)
   {Cells from the root cap / meristem just behind the cap} are actively dividing to replace those worn away (1)
   Cells furthest away from the root cap are undergoing growth and repair less (1)
   Any 2

v) 11.3/120 = 0.094 = 2.256 hours (1)
   2 hrs = 120 mins + 0.256 hr = 15.36 = 135 mins (1)

Q3
a) **Homologous chromosomes** are the same size and shape and carry the same genes but these may be different versions called alleles. (1)
   Polyploidy: where an organism has more than two complete sets of chromosomes. (1)
   Cytokinesis: the division of the cytoplasm to form two daughter cells following mitosis. (1)
   Allele: a different form of the same gene. (1)
   Interphase: a period of synthesis and growth during the cell cycle. (1)

b) Mitosis involves one cell division whereas meiosis involves two (1)
   Mitosis produces genetically identical cells whereas meiosis produces genetically different cells (1)
   In mitosis cells are diploid, whereas in meiosis they are haploid (1)
   Crossing over does not occur in mitosis but it does in prophase I of meiosis (1)
   Mitosis does not involve independent assortment whereas meiosis does (1)
   Any 4

c) i)   There are two cell divisions rather than one (1)
        the final cell produced contains half the DNA of the parent cell (1)

  ii)   Ref meiosis I – homologous chromosomes separate and each one pulled to either pole (1)
        during anaphase I as spindle fibres shorten (1)

  iii)   Each chromosome exists as two sister (identical) chromatids (1)
         chromatids then separate into chromosomes which are pulled to each pole (1)
         During anaphase II when spindle fibres shorten (1)

Q4

You are not awarded a tick per point, but rather an assessment of your answer is made awarding three main bands. Awarding a mark within the band will depend on how fully you meet the statement.

7–9 marks
Indicative content of this level is...
Fully labelled diagrams of key stages involved
Detailed description of significances of meiosis
Detailed description of steps meiosis I and meiosis II
Detailed explanation of how steps achieve significances described

The candidate constructs an articulate, integrated account, correctly linking relevant points, such as those in the indicative content, which shows sequential reasoning. The answer fully addresses the question with no irrelevant inclusions or significant omissions. The candidate uses scientific conventions and vocabulary appropriately and accurately.

4–6 marks
Indicative content of this level is...
Partially labelled diagrams of key stages involved
Description of significances of meiosis
Description of steps meiosis I and meiosis II
Explanation of how steps achieve significances described

The candidate constructs an account correctly linking some relevant points, such as those in the indicative content, showing some reasoning. The answer addresses the question with some omissions. The candidate usually uses scientific conventions and vocabulary appropriately and accurately.

1–3 marks
Indicative content of this level is...
Diagrams incomplete or some missing
Basic description or only one significance of meiosis detailed
Basic description of steps meiosis I and meiosis II
Basic explanation of how steps achieve significance(s) described or steps not linked

The candidate makes some relevant points, such as those in the indicative content, showing limited reasoning. The answer addresses the question with significant omissions. The candidate has limited use of scientific conventions and vocabulary.

0 marks
The candidate does not make any attempt or give a relevant answer worthy of credit.

A good answer would therefore include:
- Generating genetic variation through crossing over (prophase 1) and independent assortment (metaphase 1 and 2).
- Details of key events in each stage identified and how crossing over and independent assortment are brought about.
- Keeping the chromosome number constant by producing haploid gametes that recombine during fertilisation, restoring the diploid number in the zygote.
- Reference to stages where chromosome number is halved, i.e. cytokinesis following meiosis I and meiosis II.

Practice questions:
Component 2: Biodiversity and Physiology of Body Systems

Section 1: Classification and biodiversity

Q1 a) i) $21 \times 18 / 9 = 42$

     ii)   To eliminate sampling bias

     iii)   Marking must not harm or make beetles more visible to predators (1)
           time must be allowed for beetles to reintegrate into population – 24 hrs (1)

     iv)   No births, deaths, immigration or emigration has occurred between collecting samples (1)

Q2 a) Choose fields with similar abiotic factors, e.g. moisture, sunlight (1)
     choose fields of a similar size (1)
     sampling bias – quadrats should be randomly placed (1)
     Any 2

b)

Species	Field A Number of plants per m² in recently farmed field (n_1)	$n_1(n_1-1)$	Field B Number of plants per m² in unfarmed field (n_2)	$n_2(n_2-1)$
Buttercup	1	**0**	12	**12 × 11 = 132**
Daisy	3	**6**	8	**8 × 7 = 56**
Plantain	0	**0**	9	**9 × 8 = 72**
Clover	0	**0**	13	**13 × 12 = 156**
Thistle	0	**0**	12	**12 × 11 = 132**
Dandelion	1	**0**	11	**11 × 10 = 110**
Nettle	1	**0**	0	**0**
	N = **6** N−1= **5**	Σ$n_1(n_1-1)$ = **6**	N = **65** N−1= **64**	Σ $n_2(n_2-1)$ = **658**

1 mark for $n_1(n_1-1)$ and $n_2(n_2-1)$ columns correct (2)

Field A

$$S = 1 - \frac{6}{6 \times 5}$$

   = 1 − 0.2 (1) for correct working

   = <u>0.80</u> (2dp) (1) for answer

Field B

$$S = 1 - \frac{658}{65 \times 64}$$

   = 1 − 0.16 (2dp) (1) for correct working

   = <u>0.84</u> (1) for answer

c) Increase the number of quadrats used (1)
   repeat the experiment with other fields (1)

d) Species diversity increases from year 1 to year 5 when left to lie fallow (1)

Q3 a) Chordata/chordates accept vertebrates/vertebrata (1)

    b) D (1)

    c) Ref to all having pentadactyl limb (1)
       similar structures but very different functions (1)

    d) Evidence where structure has evolved from a common ancestor to perform a different function (1)

    e) i) phylogenetic tree (1)

        ii) hair – mammals (1)
           lungs – need all amphibians, mammals, turtles, lizards, snakes, crocodiles and birds (1)

Q4

You are not awarded a tick per point, but rather an assessment of your answer is made awarding three main bands. Awarding a mark within the band will depend on how fully you meet the statement.

7–9 marks
Indicative content of this level is...
Detailed description of natural selection
Detailed explanation of how all three adaptations might have evolved

The candidate constructs an articulate, integrated account, correctly linking relevant points, such as those in the indicative content, which shows sequential reasoning. The answer fully addresses the question with no irrelevant inclusions or significant omissions. The candidate uses scientific conventions and vocabulary appropriately and accurately.

4–6 marks
Indicative content of this level is...
Description of natural selection
Explanation of how all three adaptations might have evolved

The candidate constructs an account correctly linking some relevant points, such as those in the indicative content, showing some reasoning. The answer addresses the question with some omissions. The candidate usually uses scientific conventions and vocabulary appropriately and accurately.

1–3 marks
Indicative content of this level is...
Basic description of natural selection
Brief explanation of how all three adaptations might have evolved or only 1 or 2 covered

The candidate makes some relevant points, such as those in the indicative content, showing limited reasoning. The answer addresses the question with significant omissions. The candidate has limited use of scientific conventions and vocabulary.

0 marks
The candidate does not make any attempt or give a relevant answer worthy of credit.

A good answer would therefore include:

Evolution is the process by which new species are formed from pre-existing ones over a period of time. Darwin's observations of variation within a population led to the development of natural selection:

- Darwin recognised that species changed
- Proposing the theory of natural selection to explain why it happened
- Organisms overproduce offspring
- So that there is a large variation of genotypes in population
- Changes to environmental conditions bring new selection pressures through competition/predation/disease
- Only those individuals with beneficial alleles conferring a beneficial phenotype have a selective advantage, e.g. white fur in arctic, therefore are more likely to survive

- These individuals then reproduce
- Offspring are likely to inherit the beneficial alleles
- Therefore the beneficial allele frequency increases within the gene pool

Adaptations: any good example accepted so long as each type is covered with advantages, e.g.
- Anatomical, e.g. beak shape in finches able to exploit different foods
- Physiological, e.g. haemoglobin with higher affinity for oxygen allowing llamas to live at high altitude
- Behavioural, e.g. nocturnal animals avoiding heat of the day

Section 2:
Adaptations for gas exchange

Q1 a) Thin {wall of alveolus/squamous epithelial cell} reducing diffusion distance (1)
Surrounded by capillaries {to reduce diffusion distance/provides good blood supply} (1)
fluid containing surfactant provides moist surface so gases can dissolve (1)

b) DPCC molecules reduces surface tension between fluid molecules (1)
which reduces cohesive forces (1)
stabilising alveolar wall (1)
preventing alveolar wall from collapsing and sticking together (1)

Q2 a) Counter-current flow (1)

b) (in bony fish) Blood and water flow in opposite directions unlike in cartilaginous fish in which it flows in the same direction (1)
this ensures that there is always a higher concentration of oxygen in the water than the blood it meets in bony fish (1)
so diffusion is maintained along the entire length of the gill lamellae unlike cartilaginous fish (1)
so an equilibrium is not reached at 50% and more oxygen can be extracted from the water (1)
comparison needed

c) {They dry out/lose water/no longer remain most} so gases cannot dissolve (1)
{They clump together/collapse} so surface area is reduced (1)

Q3 a) The evaporation of water vapour from the leaves or other above-ground parts of the plant out through stomata into the atmosphere (1)

b) i) Graph with title and axis labelled (1)
all plots correct (2) −1 mark for each incorrect plot
Axis correct with origin (1), i.e. potassium ion concentration on horizontal axis
Both lines drawn point to point (1)
−1 if scale poor, i.e. not using at least ¾ of paper

ii) As concentration of potassium ions increases so does concentration of malate ions (1)
Potassium ions are actively transported into guard cells which triggers starch to be converted into malate (1)
This lowers the water potential of the guard cells as malate is soluble (1)
water enters by osmosis (1)
guard cells expand as the thinner outer wall can expand more than thicker inner wall so {creating a pore / causing more stomata to open} (1)
Any 4

iii) Control variables, e.g. ensure same light intensity/same temperature/same age of plants/same height of plants
ANY 2
NOT reference to repeats

Q4 a) Moist to allow gases to dissolve (1)
thin to provide a short diffusion distance (1)
permeable to allow gases to diffuse (1)

b) **Either** good blood supply to maintain the concentration gradient (1)
or ventilation mechanism to maintain concentration gradient (1)
MAX 1

Q5

You are not awarded a tick per point, but rather an assessment of your answer is made awarding three main bands. Awarding a mark within the band will depend on how fully you meet the statement.

7–9 marks
Indicative content of this level is...
Detailed description of mammals, frogs, reptiles/birds and insect gas exchange mechanisms
Detailed explanation of how each is adapted to its environment

The candidate constructs an articulate, integrated account, correctly linking relevant points, such as those in the indicative content, which shows sequential reasoning. The answer fully addresses the question with no irrelevant inclusions or significant omissions. The candidate uses scientific conventions and vocabulary appropriately and accurately.

4–6 marks
Indicative content of this level is...
Description of two/three terrestrial animals from mammals, frogs, reptiles/birds and insect gas exchange mechanisms
Explanation of how each is adapted to its environment

The candidate constructs an account correctly linking some relevant points, such as those in the indicative content, showing some reasoning. The answer addresses the question with some omissions. The candidate usually uses scientific conventions and vocabulary appropriately and accurately.

1–3 marks
Indicative content of this level is...
Basic description of two terrestrial animals from mammals, frogs, reptiles/birds and insect gas exchange mechanisms
Basic explanation of how each is adapted to its environment

The candidate makes some relevant points, such as those in the indicative content, showing limited reasoning. The answer addresses the question with significant omissions. The candidate has limited use of scientific conventions and vocabulary.

0 marks
The candidate does not make any attempt or give a relevant answer worthy of credit.

A good answer would therefore include:
- Mammals have lungs, ref to large S.A., moist surface to aid diffusion, thin walls providing short diffusion pathway, circulatory system with haemoglobin to maintain concentration gradient, internal lungs to minimise water loss, ventilation mechanism ensures fresh oxygen is brought to the exchange surface and carbon dioxide removed to maintain concentration gradient.
- Frogs can use skin for gas exchange when inactive, but can use lungs when more active, tadpoles have gills as they live in water. Ref to lung structure and function as for mammals.
- Reptiles/birds have more efficient lungs than amphibians, and air sacs which act as bellows. Lungs similar structure/function as above.
- Insects have a tracheal system lined with chitin to reduce water loss and openings called spiracles to further reduce water loss. Gas is exchanged directly with tissues as there is no blood or pigment like haemoglobin.

Section 3:
Adaptations for transport

 You are not awarded a tick per point, but rather an assessment of your answer is made awarding three main bands. Awarding a mark within the band will depend on how fully you meet the statement.

7–9 marks
Indicative content of this level is...
Detailed description of structure of arteries, capillaries and veins including fully labelled diagrams
Detailed explanation of how structure is linked to function

The candidate constructs an articulate, integrated account, correctly linking relevant points, such as those in the indicative content, which shows sequential reasoning. The answer fully addresses the question with no irrelevant inclusions or significant omissions. The candidate uses scientific conventions and vocabulary appropriately and accurately.

4–6 marks
Indicative content of this level is...
Description of structure of arteries, capillaries and veins with partially labelled diagrams or some diagrams missing
Some explanation of how structure is linked to function

The candidate constructs an account correctly linking some relevant points, such as those in the indicative content, showing some reasoning. The answer addresses the question with some omissions. The candidate usually uses scientific conventions and vocabulary appropriately and accurately.

1–3 marks
Indicative content of this level is...
Basic description of structure of arteries, capillaries and veins, diagrams may be absent or of poor quality
Structure not clearly linked to function

The candidate makes some relevant points, such as those in the indicative content, showing limited reasoning. The answer addresses the question with significant omissions. The candidate has limited use of scientific conventions and vocabulary.

0 marks
The candidate does not make any attempt or give a relevant answer worthy of credit.

A good answer would therefore include:
- Arteries have thick walls to resist high blood pressure, and elastic fibres stretch to allow the arteries to accommodate blood and the elastic recoil of the fibres pushes blood along the artery.

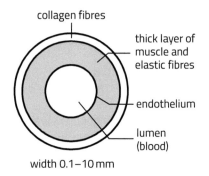

collagen fibres

thick layer of muscle and elastic fibres

endothelium

lumen (blood)

width 0.1–10 mm

- Capillaries have a narrow lumen (8–10 mm in diameter), but their total cross-sectional area is very large. Thin walls of endothelium and pores allow oxygen and nutrients to be supplied to tissues and carbon dioxide and waste to be absorbed.

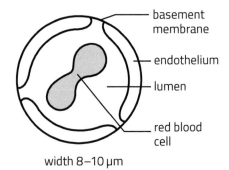

basement membrane

endothelium

lumen

red blood cell

width 8–10 μm

- Veins carry blood back to the heart. To facilitate this, they have a large diameter up to 20 mm which facilitates return of blood to the heart under low pressure, and semi-lunar valves, which prevent the backflow of blood ensuring that blood travels in one direction only.

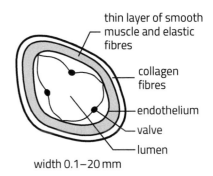

width 0.1–20 mm

- Ref to smooth endothelium in all which reduces friction of blood flow.

Q2

a) R to next R = 0.8 s (1)
Rate = 60/0.8 = 75 bpm (1)
Allow any suitable calculation from graph, e.g. P wave to P wave

b)

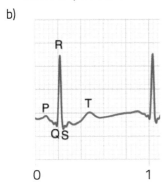

all correct for 1 mark

c) QRS = depolarisation and contraction of ventricles (1)
wave spreads from {atrioventricular node/AVN} through Bundle of His to apex of ventricle (1)
wave carried from base of ventricle upwards through the ventricle muscle causing it to contract from base upwards (1)

d) Isoelectric line labelled following T wave before next P wave (1)
represents the filling time/atria fill with blood (1)

Q3

a) Part A (is the steep part of the curve) represents co-operative binding of haemoglobin and oxygen where the haemoglobin molecule changes slightly making it easier for an additional oxygen molecule to bind (1)
This results in a rapid release of oxygen to the respiring tissues with only a small decrease in partial pressure of oxygen (1)
The fourth (and final) molecule of oxygen is more difficult to bind to haemoglobin so a large increase in partial pressure of oxygen is needed to reach full saturation part B (1)
Consequently, oxyhaemoglobin doesn't release oxygen initially until partial pressures of oxygen drop considerably (1)

b) 49% is released i.e. $0.49 \times 280 \times 10^6$ (1)
$137.2 \times 10^6 \times 4$ (as each haemoglobin can carry 4 oxygen molecules) (1)
548.8×10^6 (1)

c) i) Curve drawn to the left of normal AND labelled (1)

ii) In Foetal haemoglobin has a higher affinity for oxygen than the maternal haemoglobin (1)
and so is more saturated with oxygen at the same partial pressure of oxygen (1)
and therefore able to absorb oxygen from the mother's blood via the placenta (1)

d) i) Curve drawn to the right of normal AND labelled (1)

ii) Carbon dioxide diffuses into the red blood cell (1)
Carbonic anhydrase catalyses the reaction between carbon dioxide and water forming carbonic acid (1)
Carbonic acid dissociates into HCO_3^- and H^+ ions (1)
HCO_3^- diffuses out of the red blood cell and Cl^- ions diffuse into the cell to maintain the electrochemical neutrality / ref to the chloride shift (1)
H^+ ions combine with oxyhaemoglobin forming haemoglobinic acid (HHb)
and releasing oxygen (1)

Q4 a) Hydrostatic pressure created by blood pressure at the arteriole end forces water, salts, glucose, amino acids (and dissolved oxygen) out of the capillaries through pores in their walls (1)
Plasma proteins are too large to leave so remain in blood (1)
The water potential of the blood is lower than that of the tissue fluid (1)
tending to draw water into the capillary (1)
ref to lower hydrostatic pressure so net movement is out of the capillaries (1)
ANY 4

b) Malnourished people have lower concentrations of plasma proteins (1)
and so reabsorb less water by osmosis at the venule end of the capillary (1)
because the osmotic pressure is lower (1)

Q5 a) Answer = 2 (1)
pressure in ventricle rises above that of aorta so blood flows into aorta, opening valve (1)

b) Answer = 1 (1)
blood pressure in ventricle is higher than atrium so blood tries to pass into atrium, closing valve (1)

c) Semi-lunar valve closes preventing backflow of blood into ventricle (1)
elastic recoil of wall of aorta maintains pressure (1)
pressure rises again after 0.8s following commencement of another cycle/heartbeat/ contraction of ventricle (1)

Q6 a) Hydrophyte (1)
b) Feature correctly labelled (1) and explanation (1) from:
large air space (1)
provides a reservoir of oxygen and carbon dioxide which provides buoyancy (1)
stomata on upper surface (1)
which allows oxygen to enter leaf and carbon dioxide to leave (1)
little or no cuticle (1)
no need to prevent water loss (1)

Q7

You are not awarded a tick per point, but rather an assessment of your answer is made awarding three main bands. Awarding a mark within the band will depend on how fully you meet the statement.

7–9 marks
Indicative content of this level is...
Detailed description of structure of cortex cells and all three pathways
Detailed explanation of how structure of cortex cells aids/prevents all three pathways

The candidate constructs an articulate, integrated account, correctly linking relevant points, such as those in the indicative content, which shows sequential reasoning. The answer fully addresses the question with no irrelevant inclusions or significant omissions. The candidate uses scientific conventions and vocabulary appropriately and accurately.

4–6 marks
Indicative content of this level is...
Description of structure of cortex cells and all three pathways
Explanation of how structure of cortex cells aids/prevents all three pathways

The candidate constructs an account correctly linking some relevant points, such as those in the indicative content, showing some reasoning. The answer addresses the question with some omissions. The candidate usually uses scientific conventions and vocabulary appropriately and accurately.

1–3 marks
Indicative content of this level is...
Basic description of structure of cortex cells. Only one or two pathways described.
Basic explanation of how structure of cortex cells aids/prevents pathways described

The candidate makes some relevant points, such as those in the indicative content, showing limited reasoning. The answer addresses the question with significant omissions. The candidate has limited use of scientific conventions and vocabulary.

0 marks
The candidate does not make any attempt or give a relevant answer worthy of credit.

A good answer would therefore include:
- Three pathways described.
- Apoplast being most significant route involving water moving between the spaces in the cellulose cell wall.
- Symplast pathway – water moves through the cytoplasm and plasmodesmata (strands of cytoplasm through cell wall pits).
- Vacuolar pathway – is a minor route and involves water passing from vacuole to vacuole.
- The presence of lignin in the cell walls of the xylem vessels waterproofs them and prevents water from entering the xylem via the apoplast pathway.
- In the root, the pericycle is surrounded by a single layer of cells called the endodermis, which forms a ring surrounding the vascular tissue in the centre of the root.
- The cell walls of the endodermis are impregnated with suberin, forming an impermeable band known as the Casparian strip that drives water from the apoplast pathway into the cytoplasm.
- The endodermis helps to regulate the movement of water, ions and hormones into and out of the xylem.
- The water potential of endodermal cells is raised by water being forced into them by the Casparian strip and the active transport of sodium ions into the xylem. This lowers the water potential of fluid in the xylem, forcing water into the xylem by osmosis: this is known as root pressure.

Q8 a) ^{14}C has been incorporated into sucrose/solutes (1)
which has been transported in the phloem (1)

b) light intensity (1)
temperature (1)
carbon dioxide concentration (1)
allow water availability (1)
ANY 2

Q9 a)

Temperature /°C	Distance bubble moved in one minute / cm	Distance bubble moved in one minute / cm	Distance bubble moved in one minute / cm	Average distance bubble moved in one minute / cm	Average volume water lost mm³ per min
10	1.1	1.3	0.9	**1.1**	**8.6**
20	2.3	2.1	2.3	**2.2**	**17.2**
30	4.1	4.5	4.7	**4.4**	**34.5**
40	8.9	8.3	8.1	**8.4**	**65.9**
50	7.2	6.8	7.0	**7.0**	**55.0 (acc 54.9)**

1 mark for each column correct rounding to 1 dp
1 mark for method, i.e. converting cm to mm by × 10 then × 3.14 × radius² × distance

b) Cut stem and fit to potometer underwater to prevent air bubbles in xylem (1)
seal all joints with Vaseline to prevent air entry (1)
blot leaves dry to prevent humid layer on leaf surface (1)
ANY 2

c) Air movement using a fan at set speed/in closed box to prevent air movement
Light intensity using controlled light source and black out all other light sources

d) Increasing temperature increases the rate of transpiration (1)
doubling temperature, doubles rate of transpiration up to 40°C (1)
Above 40°C, rate of transpiration decreases (1)
ref to data (1)
ANY 3

Q10 a) **Adhesive forces** are created between charges on water molecules and their attraction with the hydrophilic lining of the vessels (1)
whereas **cohesive forces** are created by the attractive forces between water molecules due to their dipolar charges (1)

b) **Transpiration** is the evaporation of water vapour from the leaves (or other above ground parts of a plant) through stomata into atmosphere (1) whereas
capillarity is the movement of water up narrow tubes by capillary action (involving atmospheric pressure) (1)

c) **Source** is where photosynthesis occurs producing glucose (1) whereas
sink is where products of photosynthesis/glucose is used e.g. growing parts (1)

d) **Apoplast pathway** involves movement across root cortex through spaces in cellulose cell wall (1) whereas
Symplast pathway involves movement across root cortex through the cytoplasm and plasmodesmata (1)
Comparison needed

Section 4:
Adaptations for nutrition

Q1 a)

Letter	Name	Function
B	**Microvilli**	Increases surface area of B
C	**Mucosa**	Contains glands which produce mucus and enzymes
D	Submucosa	**Connective tissue containing blood and lymph vessels to remove absorbed products of digestion**
E	Circular and longitudinal muscle layers	**Contract to push food along by peristalsis**

b) Transports amino acids from the intestine to the liver (1)

c) Surface area is reduced for <u>absorption and digestion</u> (1)
so less glucose is absorbed for respiration, resulting in fatigue (1)
diarrhoea results as less water can be absorbed (1)

Q2 a) Lives inside a host organism obtaining nutrients over a period of time causing harm to the host (1)

b) Has male and female reproductive structures to allow for sexual reproduction without second tapeworm (1)
Produces vast numbers of eggs which increases chances of finding another host (1)
Eggs have resistant shells so can survive until eaten by secondary host (1)

c) Prevents synthesis of enzyme inhibitors (1)
Thins cuticle (1)

Q3 a) Autotrophic nutrition involves making own organic food from simple inorganic raw materials (1)
whereas heterotrophic nutrition involves consuming complex organic molecules produced by autotrophs (1)

b) Saprotrophic nutrition involves feeding on dead or decaying matter by secreting enzymes extracellularly and absorbing products (1)
whereas holozoic nutrition involves ingesting food and digesting before absorbing nutrients (1)

c) Photoautotrophic nutrition involves using light energy to produce organic molecules by photosynthesis (1)
whereas chemoautotrophic nutrition involves using energy from chemical reactions (1)

d) Excretion is eliminating waste made within the body e.g. urea or carbon dioxide, (1)
whereas egestion is getting rid of undigested food via faeces (1)
<u>Comparison needed</u>

Q4 You are not awarded a tick per point, but rather an assessment of your answer is made awarding three main bands. Awarding a mark within the band will depend on how fully you meet the statement.

7–9 marks
Indicative content of this level is...
Detailed description of protein digestion in all parts of the alimentary canal: mouth, stomach, duodenum and ileum
Detailed explanation of absorption of amino acids by active transport and facilitated diffusion

The candidate constructs an articulate, integrated account, correctly linking relevant points, such as those in the indicative content, which shows sequential reasoning. The answer fully addresses the question with no irrelevant inclusions or significant omissions. The candidate uses scientific conventions and vocabulary appropriately and accurately.

4–6 marks

Indicative content of this level is…

Description of protein digestion in main parts of the alimentary canal: mouth, stomach, duodenum and ileum

Explanation of absorption of amino acids by active transport

The candidate constructs an account correctly linking some relevant points, such as those in the indicative content, showing some reasoning. The answer addresses the question with some omissions. The candidate usually uses scientific conventions and vocabulary appropriately and accurately.

1–3 marks

Indicative content of this level is…

Basic description of protein digestion involving some of the key parts, e.g, mouth, and small intestine

Basic explanation of absorption of amino acids

The candidate makes some relevant points, such as those in the indicative content, showing limited reasoning. The answer addresses the question with significant omissions. The candidate has limited use of scientific conventions and vocabulary.

0 marks

The candidate does not make any attempt or give a relevant answer worthy of credit.

A good answer would therefore include:

Proteins

- Mouth – ref to mechanical digestion
- Stomach – peptidase from gastric glands – proteins to polypeptides (pH2). Ref to activation of pepsinogen to pepsin by H^+ ions in HCL from oxyntic cells in gastric pits.
- Duodenum – endopeptidases from pancreas (pH 7) – protein to polypeptides
- Ileum – endopeptidases and exopeptidases (pH 8.5) from ileum mucosa break down polypeptides to amino acids
- Amino acids absorbed in ileum by active transport into epithelial cells and then facilitated diffusion into capillary of villus
- Ref peristalsis
- Ref mucus from goblet cells lubricating food and protecting lining
- Ref highly folded ileum increasing surface area for absorption and digestion (as enzymes membrane bound)
- Ref villi and microvilli

Q5 a) Ileum (1)

b) From top to bottom, mucosa, submucosa, circular muscle, longitudinal muscle.

4 correct = 3 marks

3 correct = 2 marks

2 correct = 1 mark

c) Distance shown on image, between 42-61 mm (or allow correct measurement from image)

e.g. 61 mm

Convert to µm × 1000 = 61000 µm (1)

Size of object = 61000 / 40 = 1525 µm (1)

Allow correct calculation from image

Practice paper: Component 1 answers

1

Statement	Letter(s)
IWould be found in nucleic acids	A
May contain C=C bonds	E
Contains a glycosidic bond	B
Is a triose sugar	G

1 mark per row

2 a)

All labels correct = 2
One incorrect = 1

b) Deoxyribose in DNA vs ribose in RNA (1)
thymine in DNA vs uracil in RNA (1)
DNA very long molecule, RNA short (1)
DNA double-stranded vs usually single-stranded RNA (1)
ANY 3 but must be comparative

c) size of object = size of image / magnification
90 mm (or correct measurement from diagram) / 700,000 × 1,000,000 (1)
= 128.6 nm (1)
do not accept 0.128 mm, only 1 mark if no units

d) i) Genetic material is RNA whereas bacteria contain DNA (1)
HIV has no {cytoplasm/organelles or named organelles/chromosomes} (1)

ii) NNRTIs inhibit reverse transcriptase so prevent reverse transcription of RNA to DNA (1)
RNA cannot enter the CD4 cell nucleus and combine with the cell's nuclear material (1)

iii) NNRTIs bind reversibly so can be reused (1)
NNRTIs are non-competitive inhibitors so effect is not overcome by increasing the amount of RNA substrate (1)

3 a)

pH	Time taken for solution to become translucent/s				Rate of reaction $/ s^{-1} \times 10^{-3}$
	Trial 1	Trial 2	Trial 3	Mean	Mean
6	433	436	420	**429.67**	**2.33**
7	201	220	156	**192.33**	**5.20**
8	130	119	150	**133.00**	**7.52**
9	91	120	92	**101.00**	**9.90**
10	358	355	378	**363.67**	**2.75**

1 mark each correct column – accept nearest whole second

b)   Graph:
- correct axes labelled including units (1)
- suitable linear scale used on each axis including a value for the origin (1)
- accurate plotting of points (2) (1mark lost for each error)
- suitable and accurate joining of points with no extrapolation (1)

c)   The optimum pH = 9 as this returned the highest rate of reaction at 9.90 s^{-1} × 10^{-3} (1)
at pH 10 enzyme is denaturing as the rate drops significantly to 2.75 s^{-1} × 10^{-3} (1)
but at pH 8 the rate has only dropped to 7.52 s^{-1} × 10^{-3} (1) suggesting the enzyme is starting to inactivate (1)
the rate of reaction continue to fall as pH decreases, due to the enzyme denaturing (1)
ANY 3

d)   Results show a large range at some pHs, e.g. pH 7 where time taken ranges from 156–220 seconds (1)
measuring time taken to turn translucent is subjective (1)
use a colorimeter to record transparency after a set period of time, e.g. 30 seconds (1)

4   a)   Phospholipid bilayer correctly drawn (1)
with hydrophobic tails (1) and hydrophilic heads (1) labelled

b)   Triglyceride consists of three fatty acids and glycerol (1)
whereas in a phospholipid one fatty acid is replaced by a phosphate molecule (1)

c)   Membrane is called fluid mosaic model (1)
phospholipids are free to move (within membrane/bilayer) (1)
and hence proteins are free to move within {membrane/bilayer} (1)
creating new arrangement of proteins (after 1 hour) (1)
mouse and human cells have same membrane structure (as able to fuse) (1)
ANY 4

5   a)   During {metaphase/drawing C} no crossing over is seen (1)
comparing prophase/D with cytokinesis/A chromosome number remains the same (1)
no pairing of homologous chromosomes seen (1)

b)   A = cytokinesis (not telophase as cell starting to cleave)
B = anaphase
C = metaphase
D = prophase
4 correct = 3, 3 correct = 2, 2 correct = 1

c)   Spindle fibres shorten (1)
centromeres divide (1)
chromatids pulled to either pole/opposite ends of cell (1)
credit ref to ATP required to remove tubulin from spindle fibres (1)
MAX 3

6   a)   In total there are 26 cells of which 15 are plasmolysed (1)
calculation of % cells plasmolysed 15/26 × 100 = 57.7% acc. 58% (1)
this is greater than 50% plasmolysed so molarity must be less than 0.5M

ii)   Repeat the experiment with range of different molarity sucrose solutions and add a known volume of sucrose solution, e.g. 1 cm^3, rather than 5 drops (1)
count more cells to improve reliability, e.g. 50 cells (1)
draw a graph of % plasmolysis against sucrose concentration /M to find 50% plasmolysis (1)

At this point water potential = solute potential (as pressure potential = 0) (1)
Convert molarity of sucrose solution to solute potential kPa using a table/known values (1)
MAX 4

7 a) The triple of bases in {mRNA/DNA} (1)
which codes for a particular amino acid or punctuation signal (1)

b) All correct = 3, one incorrect = 2, two incorrect = 1

Sequence for Sickle-Cell Haemoglobin												
ATG	GTG	CAC	CTG	ACT	CCT	GTG	GAG	AAG	TCT	GCC	GTT	ACT
START	Val	**His**	**Leu**	**Thr**	**Pro**	**Val**	**Glu**	**Lys**	**Ser**	**Ala**	**Val**	**Thr**

c) 6th Glu to Val (1)

d) Introns are non-coding nucleotide sequences found in DNA and mRNA (1)
whereas exons are the nucleotide sequence on one strand of the DNA molecule where the
corresponding mRNA codes for the production of a specific polypeptide (1)
introns are only found in eukaryotes /not in prokaryotes (1)
pre-mRNA is spliced to remove {introns/non-coding regions} before passing to the ribosomes (1)
ref to spliceosome/enzymes (1)
need comparison

8 You are not awarded a tick per point, but rather an assessment of your answer is made awarding three
main bands. Awarding a mark within the band will depend on how fully you meet the statement.

7–9 marks
Indicative content of this level is...
Detailed comparison of structure of prokaryotes, mitochondria and chloroplasts
Evaluation of how evidence supports theory

*The candidate constructs an articulate, integrated account, correctly linking relevant points, such as those
in the indicative content, which shows sequential reasoning. The answer fully addresses the question with
no irrelevant inclusions or significant omissions. The candidate uses scientific conventions and vocabulary
appropriately and accurately.*

4–6 marks
Indicative content of this level is...
Comparison of structure of prokaryotes, mitochondria and chloroplasts
Some evaluation of how evidence supports theory

*The candidate constructs an account correctly linking some relevant points, such as those in the indicative
content, showing some reasoning. The answer addresses the question with some omissions. The candidate
usually uses scientific conventions and vocabulary appropriately and accurately.*

1–3 marks
Indicative content of this level is...
Basic comparison of structure of prokaryotes, mitochondria and chloroplasts
Little or no evaluation of how evidence supports theory

*The candidate makes some relevant points, such as those in the indicative content, showing limited
reasoning. The answer addresses the question with significant omissions. The candidate has limited use of
scientific conventions and vocabulary.*

0 marks
The candidate does not make any attempt or give a relevant answer worthy of credit.

A good answer would therefore include:

Mitochondrion
<u>folded</u> {inner membrane/cristae}
increases surface area for attachment of enzymes
site of (aerobic) respiration/ATP production
credit for good diagram with correct labels, e.g. cristae, matrix, stalked particles, inter membrane space, DNA and 70S ribosomes
link to specialised respiratory bacteria

Chloroplast
stroma contains enzymes
thylakoids contain photosynthetic pigments
thylakoids stacked into grana
function is photosynthesis
credit for good diagram with correct labels, e.g. stroma, thylakoid, 70S ribosomes, circular DNA, starch grain, lipid droplet, lamellae
link to photosynthetic bacteria

Similar to prokaryotic cell
all three contain 70S ribosomes
are of a similar size
contain own DNA which helps with self-replication
credit for good diagram with correct labels, e.g. 70S ribosomes, circular DNA and ribosomes

Practice paper: Component 2 answers

1 a) Involves consumption of already made complex organic molecules (1)

    b)   i)   Animal 1 = carnivore (1)
                  Animal 2 = herbivore (1)

        ii)   Carnassial teeth (1)

        iii)   In animal 1 there is attachment for powerful muscles, and jaw articulates up and down only to cut through food (1)
               in animal 2 jaw moves side to side only to help grind food (1)

    c)   Animal 2 eats vegetation/plants which have a high water content (1)
        water is absorbed into the blood in the colon (1)
        so the colon is longer to absorb the higher proportion of water or WTTE (1)

2 a)   Electrocardiogram (1)

    b)   **at rest:**
        2.2 seconds = 3 beats from S–S (allow from graph)
        = 0.73 (1)
        60/0.73 = 82 beats per minute (allow 82.2) (1)

        after moderate exercise:
        1.0 seconds = 2 beats from S–S (allow from graph)
        = 0.50 (1)
        60/0.50 = 120 beats per minute (1)
        if no units −1 (don't penalise twice)
        MAX = 3

    c)   {isoelectric line/filling time/gap between T and P} decreases (1)
        to allow more beats per minute as time taken for atrial and ventricular systole is constant (1)

3 a)   Very large surface area – approx. 700 million alveoli (1)
        very thin walls approx. 0.1 μm (1)
        Surrounded by capillaries to reduce diffusion distance (1)
        Moist lining to allow gases to dissolve before diffusing (1)
        Permeable to gases (1)
        Collagen and elastic fibres allow for expansion and recoil (1)
        Any 4

    b)   Surfactant is not produced by foetus until around 23 weeks of pregnancy (1)
        surfactant lowers the surface tension (1)
        preventing the alveoli from collapsing and sticking together (1)

4 a)   Haemoglobin is a transport protein with a quaternary structure (1)
        and contains four haem groups which contain iron (1)
        oxygen binds cooperatively with each of the four haem groups (1)
        so iron deficiency leads to fewer haemoglobin molecules, and therefore less oxygen can be carried (1)
        Max 3

b) Haemoglobin has a decreased affinity for oxygen (1)
{oxygen is released more readily to the tissues / more oxygen released} at same partial pressure of oxygen (1)
Supplying more oxygen to respiring tissues (1)

c) Line drawn to the left of normal starting and finishing in same place as normal (1)
Significance:
haemoglobin has a higher affinity for oxygen than normal haemoglobin (1)
haemoglobin is more saturated at the same partial pressure of oxygen than normal (1)
and so can pick up oxygen more easily (1)

5 a) The number of species and the number of individuals of each species in a specified geographic region (1)

b) Both areas should be 'kicked' / sampled for the same period of time, e.g. 30 seconds (1)
net should be held the same distance downstream from the kick site (1)
invertebrates should be handled carefully and released carefully after identification (1)

c)

Species	Number present near factory outlet (Site A)	Site A $n(n-1)$	Number present 2 miles downstream of factory (Site B)	Site B $n(n-1)$
Stonefly nymph	0	**0**	15	**210**
Mayfly larva	0	**0**	12	**132**
Freshwater shrimp	0	**0**	11	**110**
Caddis fly larva	3	**6**	4	**12**
Bloodworm	7	**42**	2	**2**
Water louse	9	**72**	0	**0**
Red-tailed maggot	11	**110**	0	**0**
	N = **30**	$\sum n(n-1)$ = **230**	N = **44**	$\sum n(n-1)$ = **466**
	N−1 = **29**		N−1 = **43**	

Correct calculation of $\sum n(n-1)$ for site A (1)

Correct calculation of $\sum n(n-1)$ for site B (1)

Site A

$S = 1 - \dfrac{230}{30 \times 29}$

$S = 1 - \dfrac{230}{870}$

$S = 1 - 0.264 = \underline{0.74}$ (1)

Site B

$S = 1 - \dfrac{466}{44 \times 43}$

$S = 1 - \dfrac{466}{1892}$

$S = 1 - 0.246 = \underline{0.75}$ (1)

d) Simpson's diversity index returns similar {values 0.74, 0.75} for both sites so is not reliable (1)
    this is because there is not a big difference in number of species present between both sites ref to 5
    different species at site B vs 4 at site A (1)
    type of species is different between both sites, e.g.
    stonefly nymph, mayfly larva and freshwater shrimps are absent in polluted sites (1)
    water louse and red-tailed maggot are absent in unpolluted site (1)

6 a) i) Trout have an oxygen requirement of 13–14 mg/L which is only found in still water below 5°C
            which is found in typical UK winters (1)
            in summer, water temperatures are typically 8–16°C meaning dissolved oxygen saturation is
            around 10.1–11.3 mg/L which is too low for trout (1)
            Rivers are fast-moving which increases oxygen levels as air mixes into the water or WTTE (1)

       ii) Pike have a lower oxygen requirement of 3–5 mg/L (1)
            lowest dissolved oxygen saturation in a lake in summer at 15°C is 10.1 mg/L (1)

      iii) Bony fish are able to absorb more oxygen from the water due to {blood and water flowing in
            opposite directions/ref counter-current explained} (1)
            than cartilaginous fish where parallel flow occurs (1)
            as less oxygen is absorbed due to equilibrium being reached / gas exchange only occurs across
            part of gill surface (1)
            ref to bony fish having more advanced ventilation system (1)

7 a) e.g. $3.14 \times 0.25 \times 2.3 = 1.8$

Air speed / ms^{-1}	Mean distance moved by the bubble / mm	Mean volume of water taken in / mm^3 min^{-1}
0	2.3	**1.8**
5	12.0	**9.4**
10	30.3	**23.8**
15	45.0	**35.3**

All four correct = 3 marks, 3 correct = 2 marks, 2 correct = 1 mark. Credit working, allow consequential
error (i.e. don't penalise twice for same error, e.g. failure to divide diameter by 2)

   b) There is insufficient data to support the conclusion made, an additional measurement at
       20 ms^{-1} would be needed to see the effect of doubling from 10 to 20 ms^{-1} (1)
       the results 10 ms^{-1} are not reliable as they vary from 22 to 39 mm so the actual mean could lie
       between these values (1)
       need to repeat 10 ms^{-1} again (1)
       whilst the other results are much closer to each other, e.g. 5 ms^{-1} varies from 11 to 13 mm (1)
       ANY 3

   c) Ensure stem was cut and fitted underwater (1)
       blot leaves dry to prevent humid layer forming (1)
       control light intensity by blacking out all light and using one light source, e.g. a lamp at a set distance
       with a heat shield (1)
       control humidity / or ref to measuring it to see if it varied and therefore could be a factor (1)
       control temperature / or ref to measuring it to see if it varied and therefore could be a factor (1)
       ANY 4

   d) Mean volume of water intake would have been lower at all readings (1)
       marram grass has adaptations for reducing water loss (one mark for each with explanation – Max 2),
       e.g. sunken stomata which trap humid air, hairs around stomata which trap humid air, rolled leaves
       which reduces surface area, thick cuticle to reduce water loss from leaf surface

8

You are not awarded a tick per point, but rather an assessment of your answer is made awarding three main bands. Awarding a mark within the band will depend on how fully you meet the statement.

7–9 marks
Indicative content of this level is...
Detailed comparison of structure of arteries, capillaries and veins
Evaluation of how evidence supports theory

The candidate constructs an articulate, integrated account, correctly linking relevant points, such as those in the indicative content, which shows sequential reasoning. The answer fully addresses the question with no irrelevant inclusions or significant omissions. The candidate uses scientific conventions and vocabulary appropriately and accurately.

4–6 marks
Indicative content of this level is...
Comparison of structure of arteries, capillaries and veins
Some evaluation of how evidence supports theory

The candidate constructs an account correctly linking some relevant points, such as those in the indicative content, showing some reasoning. The answer addresses the question with some omissions. The candidate usually uses scientific conventions and vocabulary appropriately and accurately.

1–3 marks
Indicative content of this level is...
Basic comparison of structure of arteries, capillaries and veins
Little or no evaluation of how evidence supports theory

The candidate makes some relevant points, such as those in the indicative content, showing limited reasoning. The answer addresses the question with significant omissions. The candidate has limited use of scientific conventions and vocabulary.

0 marks
The candidate does not make any attempt or give a relevant answer worthy of credit.

A good answer would therefore include:

Arteries
Carry blood away from heart so under high pressure
Have thick walls to resist pressure
Elastic fibres to allow walls to stretch and accommodate blood
Elastic recoil helps push blood along
Vessels 0.1–10 mm wide

Capillaries
Very narrow lumen (8–10 mm)
Total cross-sectional area very large
Very thin / ref to pores in basement membrane to allow exchange of, e.g. oxygen, glucose, carbon dioxide

Veins
Carry blood back to heart under low pressure
so layer of smooth muscle and elastic fibres is thin
Vessels wide (0.1–20 mm)
Semi-lunar valves prevent back flow of blood and ensure blood travels in one direction
Surrounding skeletal muscle squeeze veins to aid return of blood